Target Grade 9

Get back on track

Edexcel GCSE (9–1)

Mathematics

Algebraic techniques, Shape and Statistics

Katherine Pate

Pearson

Published by Pearson Education Limited, 80 Strand, London, WC2R ORL.

www.pearsonschoolsandfecolleges.co.uk

Typeset by Tech-Set Ltd, Gateshead

First published 2017

19 18 17
10 9 8 7 6 5 4 3 2

British Library Cataloguing in Publication Data
A catalogue record for this book is available from the British Library

ISBN 978 0 435 18338 7

Printed in Italy by Lego S.p.A

Helping you to formulate grade predictions, apply interventions and track progress.

Any reference to indicative grades in the Pearson Target Workbooks and Pearson Progression Services is not to be used as an accurate indicator of how a student will be awarded a grade for their GCSE exams.

You have told us that mapping the Steps from the Pearson Progression Maps to indicative grades will make it simpler for you to accumulate the evidence to formulate your own grade predictions, apply any interventions and track student progress.

We're really excited about this work and its potential for helping teachers and students. It is, however, important to understand that this mapping is for guidance only to support teachers' own predictions of progress and is not an accurate predictor of grades.

Our Pearson Progression Scale is criterion referenced. If a student can perform a task or demonstrate a skill, we say they are working at a certain Step according to the criteria. Teachers can mark assessments and issue results with reference to these criteria which do not depend on the wider cohort in any given year. For GCSE exams however, all Awarding Organisations set the grade boundaries with reference to the strength of the cohort in any given year. For more information about how this works please visit: https://qualifications.pearson.com/en/support/support-topics/results-certification/understanding-marks-and-grades.html/Teacher

Each practice question features a Step icon which denotes the level of challenge aligned to the Pearson Progression Map and Scale.

To find out more about the Progression Scale for Maths and to see how it relates to indicative GCSE 9–1 grades go to www.pearsonschools.co.uk/ProgressionServices

Contents

Useful formulae

Unit 4 Histograms with unequal class widths

Frequency density $= \dfrac{\text{frequency}}{\text{class width}}$

Unit 5 Triangles

Cosine rule: $a^2 = b^2 + c^2 - 2bc \cos A$

OR $\cos A = \dfrac{b^2 + c^2 - a^2}{2bc}$

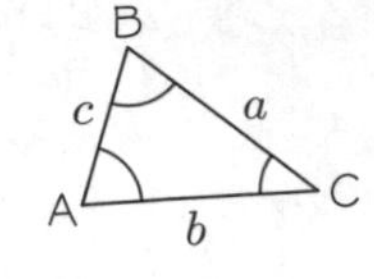

Area of a triangle $= \dfrac{1}{2}ab \sin C$

Unit 6 Pythagoras and trigonometry in 3D

Sine rule: $\dfrac{a}{\sin A} = \dfrac{b}{\sin B} = \dfrac{c}{\sin C}$

OR $\dfrac{\sin A}{a} = \dfrac{\sin B}{b} = \dfrac{\sin C}{c}$

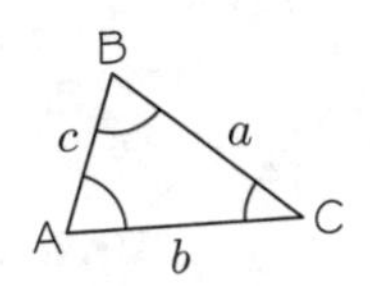

Unit 7 3D geometry

Curved surface area of a cone $= \pi rl$

Total surface area of a cone $= \pi rl + \pi r^2$

Volume of a cone $= \dfrac{1}{3}\pi r^2 h$

For two mathematically similar solids, A and B:

- length of A $= k \times$ length of B
- surface area of A $= k^2 \times$ surface area of B
- volume of A $= k^3 \times$ volume of B

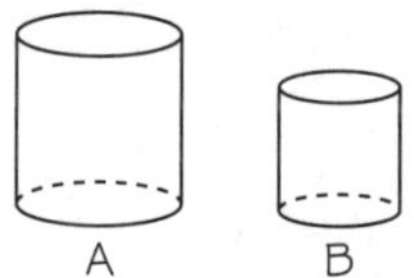

Unit 9 Circles

Area of a sector of angle θ and radius $r = \dfrac{\theta}{360} \times \pi r^2$

Arc length of a sector of angle θ and radius $r = \dfrac{\theta}{360} \times 2\pi r$

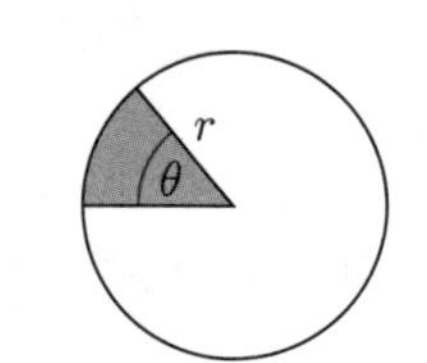

Equation of a circle, centre the origin, radius r: $x^2 + y^2 = r^2$

Gradient of a straight line $= \dfrac{\text{change in } y}{\text{change in } x}$

Glossary

Unit 1 Further algebra

Subject of a formula: the letter on its own, on one side of the equals sign.

Inverse operation: the operation that 'undoes' an operation

× is the inverse of ÷, and vice versa

'square root' is the inverse of 'square', and vice versa

Expand (brackets): multiply each term inside one bracket by the term outside the bracket or by each term in every other set of brackets.

Factorise: the reverse of expand.

Like terms: terms that have exactly the same powers of the same letters.

Variable: a letter in an algebraic expression used to represent a value that can change or vary.

Unit 2 Algebraic fractions

Numerator: the top part of a fraction.

Denominator: the bottom part of a fraction.

Factor: a number or expression that divides exactly into another number or expression.

Unit 3 Iterative processes

Iteration formula: a formula that you use repeatedly to estimate solutions to equations. You substitute the answer you obtain from the formula into the formula again until the answer is the same to the required level of accuracy.

Quadratic equation: an equation in which the highest power of x is x^2. A quadratic equation can be written in the form $y = ax^2 + bx + c$.

Cubic equation: an equation in which the highest power of x is x^3. A cubic equation can be written in the form $y = ax^3 + bx^2 + cx + d$.

Unit 4 Histograms with unequal class widths

Frequency: how many times something occurs.

Frequency density: calculated by dividing frequency by class width.

Histogram: a way of displaying data. The height of each bar is the frequency density and the area of the bar is the frequency.

Unit 6 Pythagoras and trigonometry in 3D

Pythagoras' theorem: the relationship between the lengths of the three sides of a right-angled triangle. When the two shorter sides are of lengths a and b, the length of the longest side (the hypotenuse) is given by $c^2 = a^2 + b^2$.

Diagonal of a cuboid: a straight line joining a vertex on the top face to a vertex on the bottom face, passing through the centre of the cuboid.

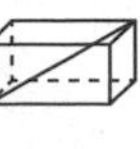

Right pyramid: a pyramid in which the apex is vertically above the centre of the base. A line from the apex to the centre of the base meets the base at 90°.

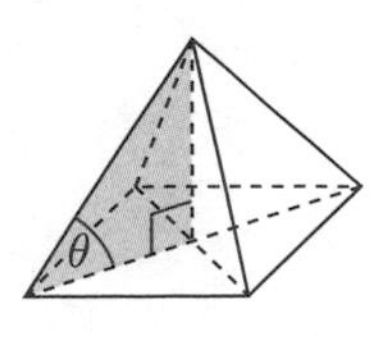

Unit 7 3D geometry

Similar shapes: when one shape is an enlargement of the other.

Linear scale factor: the ratio of lengths of similar shapes.

Area scale factor: the ratio of areas of similar shapes.

Volume scale factor: the ratio of volumes of similar shapes.

Frustum: a cone with the pointed top cut off.

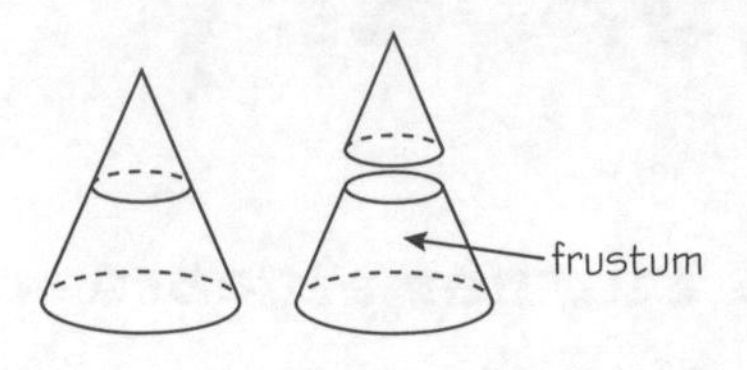

Unit 8 Algebraic and geometric proof

Consecutive: numbers that follow each other in a pattern without a gap, e.g. consecutive integers are 1, 2, 3, …; consecutive odd numbers are 1, 3, 5, …

Integer: a positive or negative whole number, or zero.

Unit 9 Circles

Radius: a straight line from the centre of a circle to the circumference; plural radii.

Tangent: a straight line that just touches a circle at one point.

Perpendicular: at 90°, at right angles to.

Sector: the area between two radii, like a slice of pie.

Chord: a straight line from one side of a circle to the other, not necessarily through the centre.

Segment: area between a chord and the circumference.

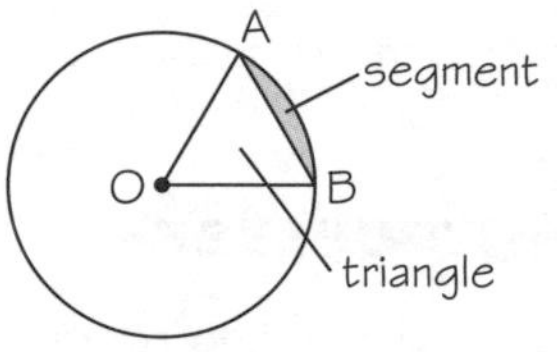

Get started

1 Further algebra

This unit will help you to rearrange formulae and expand expressions with two or three sets of brackets.

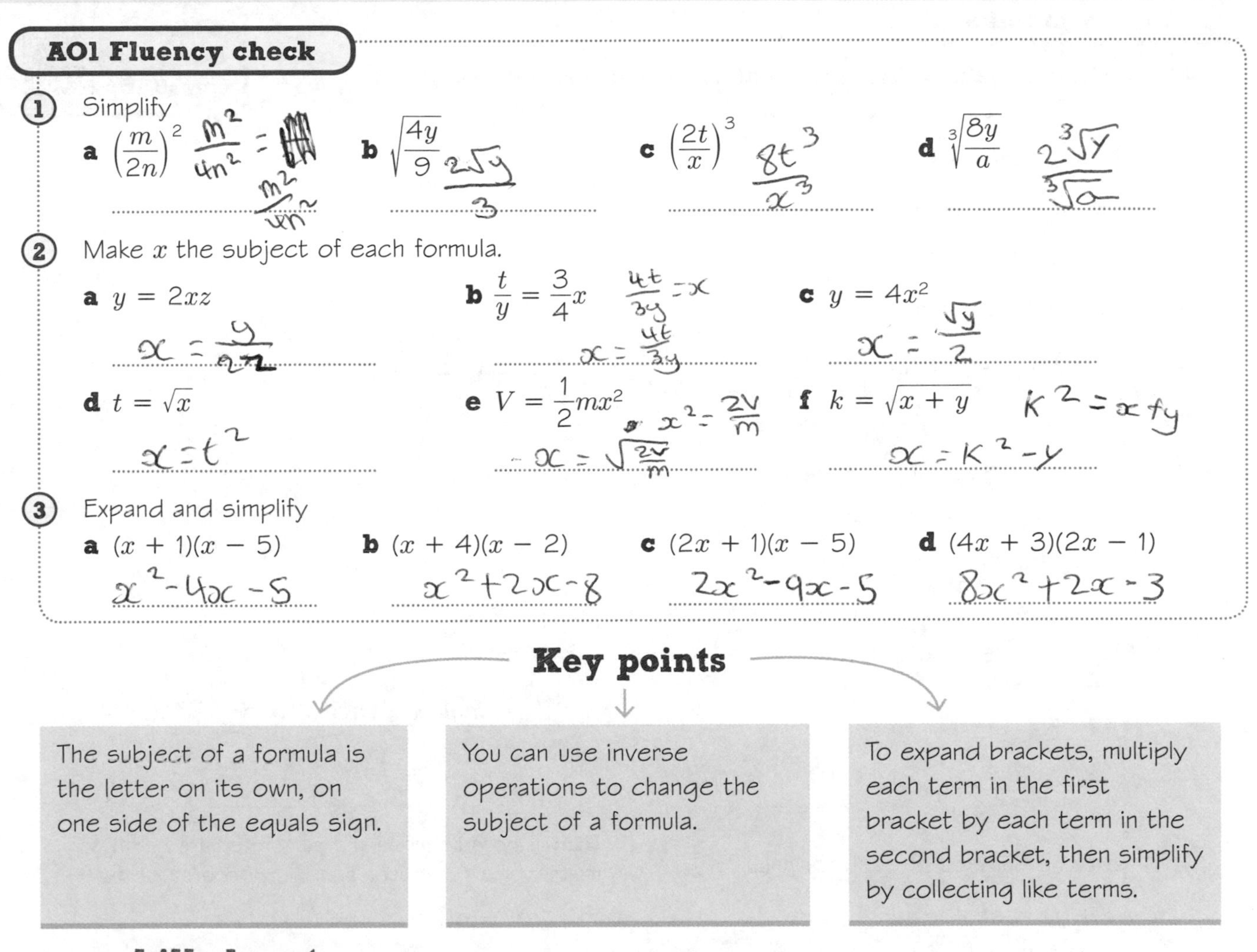

AO1 Fluency check

1. Simplify

   a $\left(\frac{m}{2n}\right)^2$ **b** $\sqrt{\frac{4y}{9}}$ **c** $\left(\frac{2t}{x}\right)^3$ **d** $\sqrt[3]{\frac{8y}{a}}$

2. Make x the subject of each formula.

   a $y = 2xz$ **b** $\frac{t}{y} = \frac{3}{4}x$ **c** $y = 4x^2$

   d $t = \sqrt{x}$ **e** $V = \frac{1}{2}mx^2$ **f** $k = \sqrt{x + y}$

3. Expand and simplify

   a $(x + 1)(x - 5)$ **b** $(x + 4)(x - 2)$ **c** $(2x + 1)(x - 5)$ **d** $(4x + 3)(2x - 1)$

Key points

The subject of a formula is the letter on its own, on one side of the equals sign.

You can use inverse operations to change the subject of a formula.

To expand brackets, multiply each term in the first bracket by each term in the second bracket, then simplify by collecting like terms.

These **skills boosts** will help you to rearrange formulae and expand expressions with two or three sets of brackets.

1 Rearranging complex formulae

2 Expanding two or three brackets

You might have already done some work on further algebra. Before starting the first skills boost, rate your confidence using each concept.

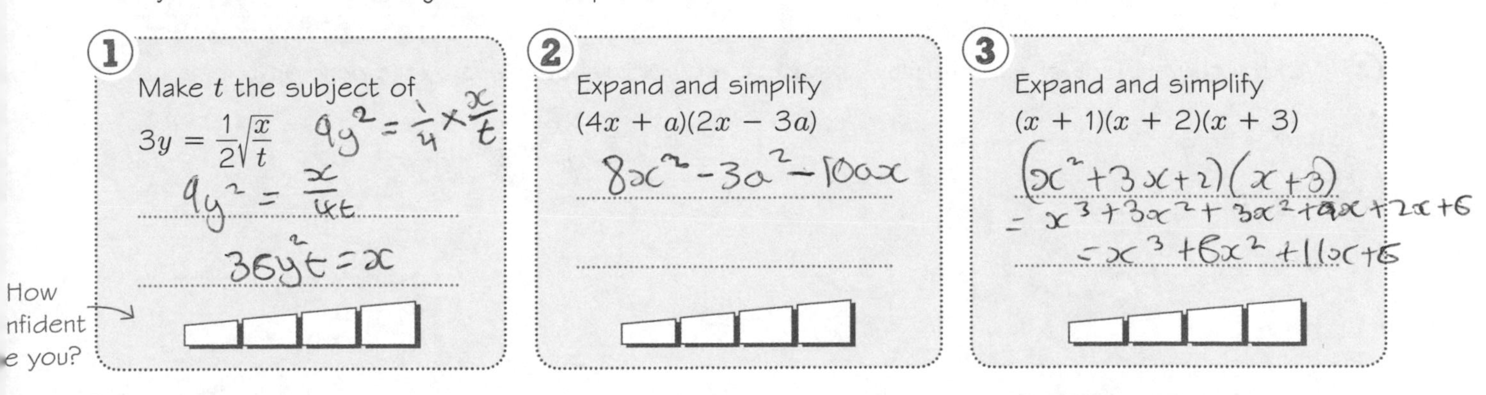

1. Make t the subject of $3y = \frac{1}{2}\sqrt{\frac{x}{t}}$

2. Expand and simplify $(4x + a)(2x - 3a)$

3. Expand and simplify $(x + 1)(x + 2)(x + 3)$

How confident are you?

1 Rearranging complex formulae

Identify the operations used on the variable: powers/roots, +, −, × or ÷.
Use the inverse operations to change the subject, doing the same to both sides.

Guided practice

Worked exam question

Make x the subject of the formula $a(bx + c) = 4x + e$

Expand the brackets.

$abx + ac = 4x + e$

Get all the x terms on one side.

$abx = 4x + e - ac$ — Subtract ac from both sides.

$abx - 4x = e - ac$ — Subtract $4x$ from both sides.

Factorise.

$x(ab - 4) = e - ac$

Divide both sides by $(ab - 4)$.

$$x = \frac{e - ac}{ab - 4}$$

1 **a** Make x the subject of $at - 5x = xy$

$at = 5x + xy$

$x = \frac{at}{5+y}$

b Make m the subject of $E = mgh + \frac{1}{2}mv^2$

$2E = 2mgh + mv^2$

$m(2gh + v^2) = 2E$ $\quad m = \frac{2E}{2gh + v^2}$

2 Make t the subject of each formula.

Hint If both terms in the numerator are negative, multiply the numerator and the denominator by −1.

a $r(t + f) = st + g$

$rt + rf = st + g$

b $2(b + at) = 3(t - c)$

Hint If both terms in the numerator have a common factor, then factorise.

c $3a(b + t) = b - 2t$

d $x - 4bt = 2ax + 3t$

3 Make y the subject of each formula.

Hint First, multiply both sides by the denominator.

a $x = \frac{y + 3}{y}$

$xy = $ ....

b $n = \frac{xy - t}{ry + 3}$

4 Make x the subject of each formula. **Hint** Get the root on its own, then square both sides.

a $m = r\sqrt{\frac{x}{a}}$

b $2y = \frac{1}{3}\sqrt{\frac{t}{x}}$

Hint Get the 'square' on its own, then square root both sides.

c $y(x + 3)^2 = m$

d $9\left(\frac{x}{a}\right)^2 = 2y$

5 Make y the subject of each formula. **Hint** Get the 'cube' on its own, then cube root both sides.

a $t = ay^3$

b $z = \frac{y^3}{8}$

c $rs = \left(\frac{y}{b}\right)^3$

d $n(y - 5)^3 = p$

e $t = \frac{(y + 2)^3}{x}$

f $\left(\frac{y}{a - 1}\right)^3 = 2b$

6 Make x the subject of each formula.

a $m = \sqrt[3]{4x}$

b $y = \sqrt[3]{\frac{x}{v}}$

c $\frac{r}{t} = a\sqrt[3]{x}$ **Hint** Get the root on its own on one side, then cube both sides.

d $y = \sqrt[3]{\frac{ax}{b}}$

e $d = \sqrt[3]{\frac{x}{y}}$

f $e = \sqrt[3]{\frac{b}{x}}$

Exam-style question

7 Make m the subject of the formula $\sqrt{\frac{2m}{5 - m}} = t$

(3 marks)

Reflect Which inverse operations have you used in rearranging these formulae?

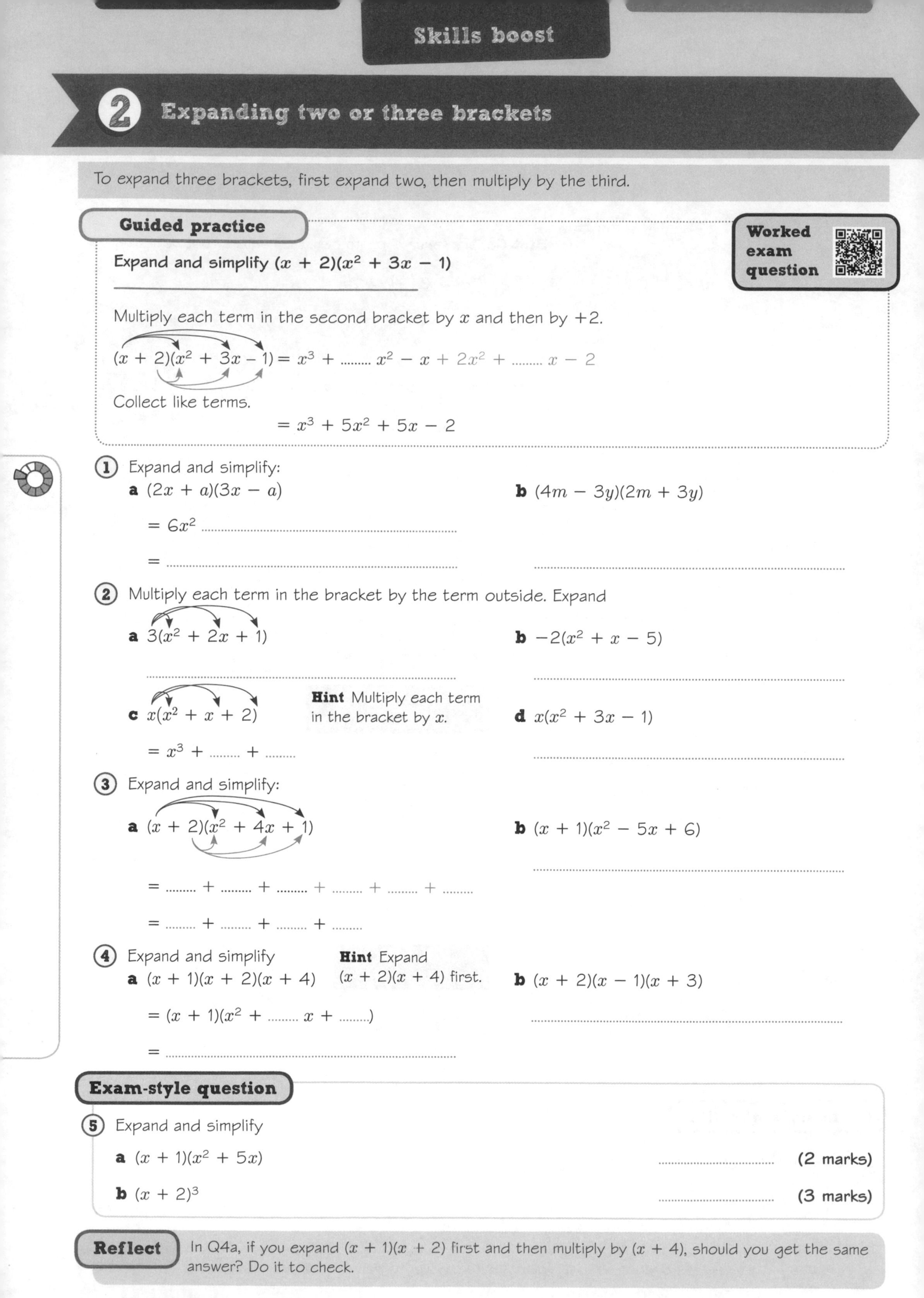

2 Expanding two or three brackets

To expand three brackets, first expand two, then multiply by the third.

Guided practice

Worked exam question

Expand and simplify $(x + 2)(x^2 + 3x - 1)$

Multiply each term in the second bracket by x and then by $+2$.

$(x + 2)(x^2 + 3x - 1) = x^3 + \dots\dots x^2 - x + 2x^2 + \dots\dots x - 2$

Collect like terms.

$= x^3 + 5x^2 + 5x - 2$

1. Expand and simplify:

   a $(2x + a)(3x - a)$

   $= 6x^2$ ....

   $=$ ....

   b $(4m - 3y)(2m + 3y)$

   ....

2. Multiply each term in the bracket by the term outside. Expand

   a $3(x^2 + 2x + 1)$

   ....

   b $-2(x^2 + x - 5)$

   ....

   c $x(x^2 + x + 2)$

   $= x^3 +$ ...... $+$ ......

   Hint Multiply each term in the bracket by x.

   d $x(x^2 + 3x - 1)$

   ....

3. Expand and simplify:

   a $(x + 2)(x^2 + 4x + 1)$

   $=$ ...... $+$ ...... $+$ ...... $+$ ...... $+$ ...... $+$ ......

   $=$ ...... $+$ ...... $+$ ...... $+$ ......

   b $(x + 1)(x^2 - 5x + 6)$

   ....

4. Expand and simplify

   a $(x + 1)(x + 2)(x + 4)$

   Hint Expand $(x + 2)(x + 4)$ first.

   $= (x + 1)(x^2 +$ ...... $x +$ ......$)$

   $=$ ....

   b $(x + 2)(x - 1)(x + 3)$

   ....

Exam-style question

5. Expand and simplify

   a $(x + 1)(x^2 + 5x)$ .... **(2 marks)**

   b $(x + 2)^3$ .... **(3 marks)**

Reflect In Q4a, if you expand $(x + 1)(x + 2)$ first and then multiply by $(x + 4)$, should you get the same answer? Do it to check.

Practise the methods

Answer this question to check where to start.

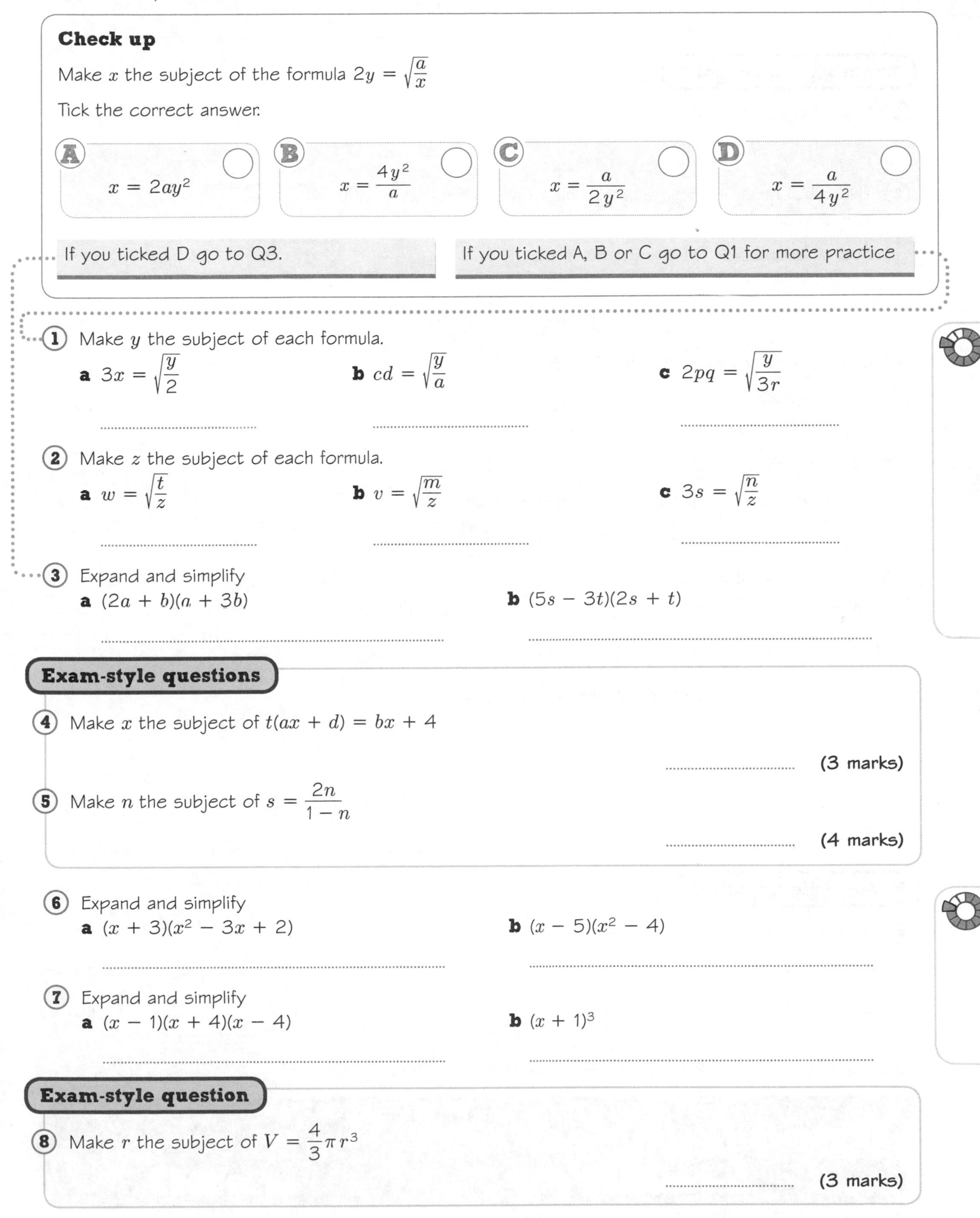

Check up

Make x the subject of the formula $2y = \sqrt{\frac{a}{x}}$

Tick the correct answer.

A $x = 2ay^2$

B $x = \frac{4y^2}{a}$

C $x = \frac{a}{2y^2}$

D $x = \frac{a}{4y^2}$

If you ticked D go to Q3.

If you ticked A, B or C go to Q1 for more practice

1 Make y the subject of each formula.

a $3x = \sqrt{\frac{y}{2}}$

b $cd = \sqrt{\frac{y}{a}}$

c $2pq = \sqrt{\frac{y}{3r}}$

2 Make z the subject of each formula.

a $w = \sqrt{\frac{t}{z}}$

b $v = \sqrt{\frac{m}{z}}$

c $3s = \sqrt{\frac{n}{z}}$

3 Expand and simplify

a $(2a + b)(a + 3b)$

b $(5s - 3t)(2s + t)$

Exam-style questions

4 Make x the subject of $t(ax + d) = bx + 4$

(3 marks)

5 Make n the subject of $s = \frac{2n}{1 - n}$

(4 marks)

6 Expand and simplify

a $(x + 3)(x^2 - 3x + 2)$

b $(x - 5)(x^2 - 4)$

7 Expand and simplify

a $(x - 1)(x + 4)(x - 4)$

b $(x + 1)^3$

Exam-style question

8 Make r the subject of $V = \frac{4}{3}\pi r^3$

(3 marks)

Problem-solve!

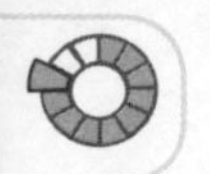

1 Write an expression for the volume of a cube of side $(x + 3)$ cm.

....

Exam-style questions

2 Make t the subject of $3(z - at) + 7 = b(t + 3)$

.... **(4 marks)**

3 This cuboid has a cuboid-shaped hole of cross-section 1 cm by 1 cm all the way through its centre.

Find an expression for the volume of the remaining solid.

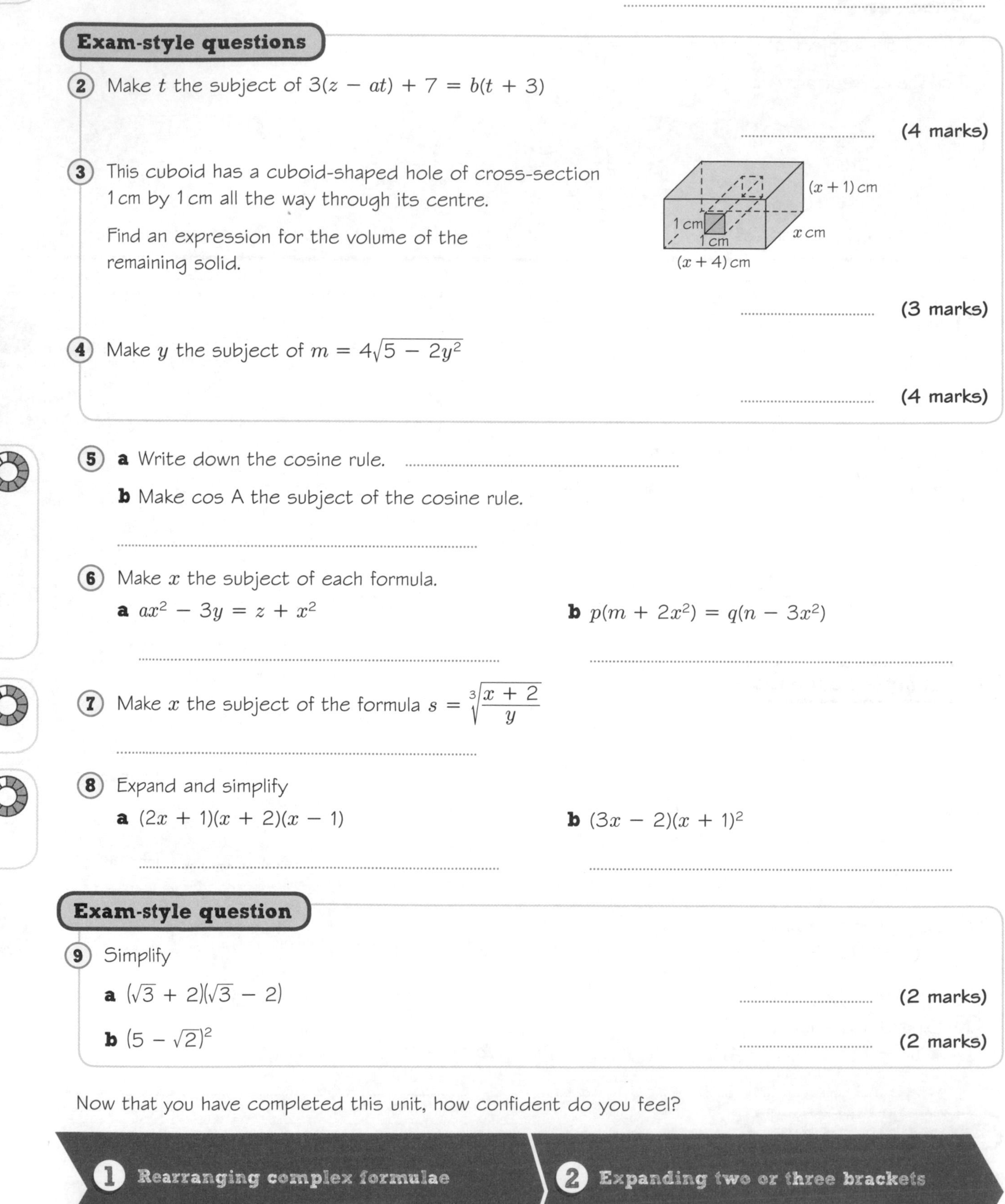

.... **(3 marks)**

4 Make y the subject of $m = 4\sqrt{5 - 2y^2}$

.... **(4 marks)**

5 **a** Write down the cosine rule. ....

b Make cos A the subject of the cosine rule.

....

6 Make x the subject of each formula.

a $ax^2 - 3y = z + x^2$

....

b $p(m + 2x^2) = q(n - 3x^2)$

....

7 Make x the subject of the formula $s = \sqrt[3]{\frac{x + 2}{y}}$

....

8 Expand and simplify

a $(2x + 1)(x + 2)(x - 1)$

....

b $(3x - 2)(x + 1)^2$

....

Exam-style question

9 Simplify

a $(\sqrt{3} + 2)(\sqrt{3} - 2)$ .... **(2 marks)**

b $(5 - \sqrt{2})^2$ .... **(2 marks)**

Now that you have completed this unit, how confident do you feel?

1 Rearranging complex formulae

2 Expanding two or three brackets

2 Algebraic fractions

This unit will help you to work with algebraic fractions and solve equations.

AO1 Fluency check

1. Factorise

   a $2x^2 - 5x$ ....

   b $x^2 - 25$ ....

   c $x^2 + 6x + 5$ ....

   d $2x^2 - x - 6$ ....

2. Simplify

   a $\frac{x}{x^2}$ ....

   b $\frac{6x^3}{x}$ ....

   c $\frac{rs}{rs^2}$ ....

   d $\frac{8x^2y^3}{6xy^2}$ ....

3. Write as a single fraction

   a $\frac{3}{5}x + \frac{x}{4}$ ....

   b $\frac{1}{2x} - \frac{1}{6x}$ ....

   c $\frac{x+1}{3} + \frac{x}{2}$ ....

4. Solve

   a $x^2 + 2x - 15 = 0$ ....

   b $2x^2 - 5x + 1 = 0$ ....

Key points

An algebraic fraction is a fraction with letter terms in the numerator or the denominator or both.

You can simplify, add, subtract, multiply and divide algebraic fractions using the same methods as with number fractions.

These **skills boosts** will help you to work with algebraic fractions and solve equations.

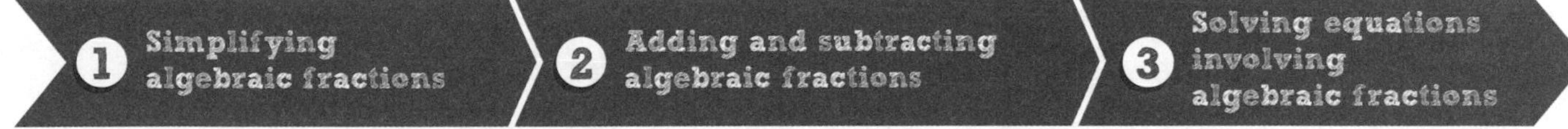

You might have already done some work on algebraic fractions. Before starting the first skills boost, rate your confidence using each concept.

1. Simplify $\frac{x^2 + 3x - 4}{x^2 + x - 12}$ ....

2. Add $\frac{1}{x-4} + \frac{1}{x+1}$ ....

3. Solve $\frac{1}{x-2} + \frac{5}{x+2} = 2$ ....

How confident are you?

1 Simplifying algebraic fractions

To simplify an algebraic fraction:

- factorise the numerator and factorise the denominator
- divide the numerator and denominator by any common factors.

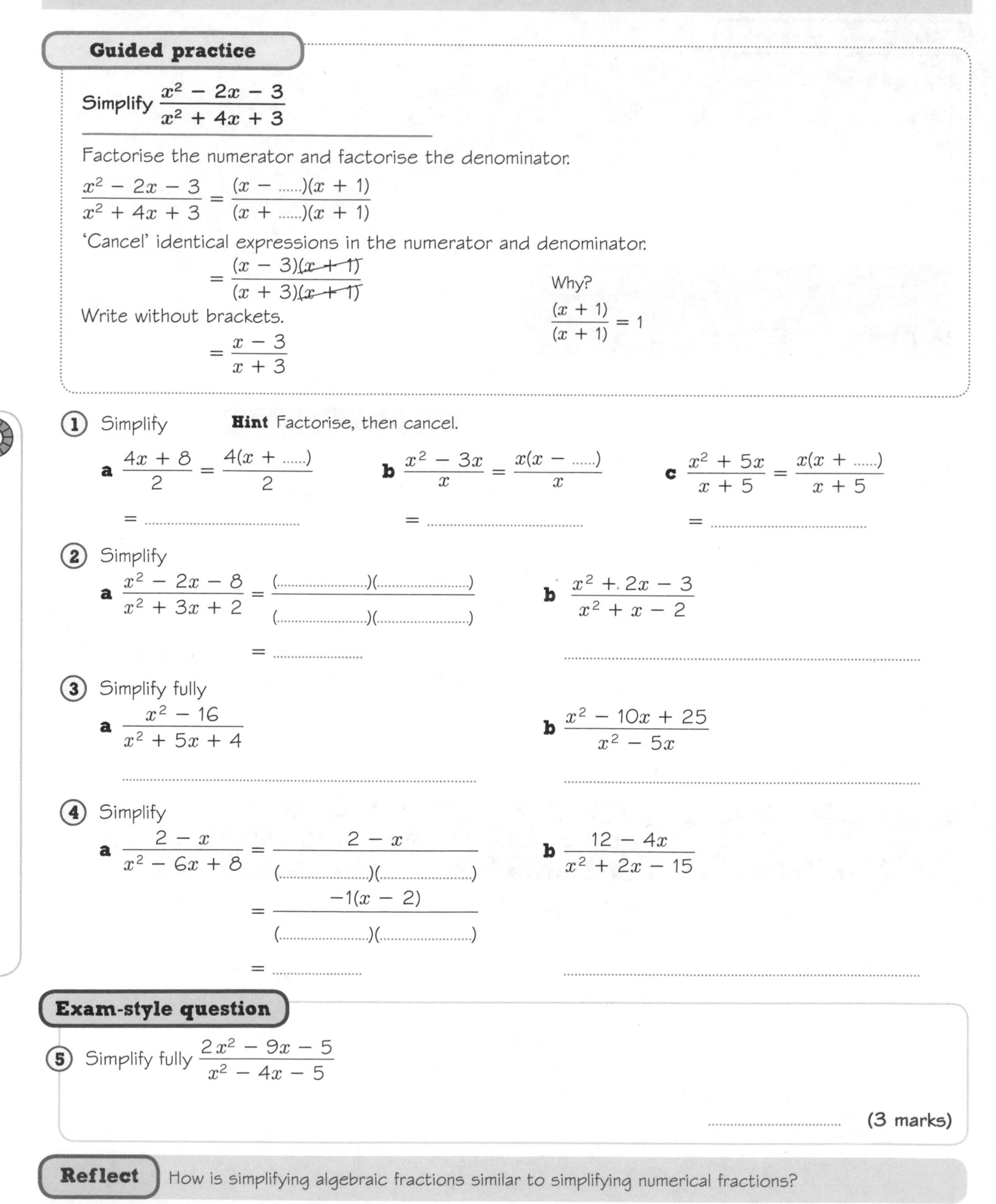

Guided practice

Simplify $\dfrac{x^2 - 2x - 3}{x^2 + 4x + 3}$

Factorise the numerator and factorise the denominator.

$$\frac{x^2 - 2x - 3}{x^2 + 4x + 3} = \frac{(x - \text{......})(x + 1)}{(x + \text{......})(x + 1)}$$

'Cancel' identical expressions in the numerator and denominator.

$$= \frac{(x - 3)\cancel{(x + 1)}}{(x + 3)\cancel{(x + 1)}}$$

Why? $\dfrac{(x + 1)}{(x + 1)} = 1$

Write without brackets.

$$= \frac{x - 3}{x + 3}$$

1 Simplify **Hint** Factorise, then cancel.

a $\dfrac{4x + 8}{2} = \dfrac{4(x + \text{......})}{2}$ $= \text{..........}$

b $\dfrac{x^2 - 3x}{x} = \dfrac{x(x - \text{......})}{x}$ $= \text{..........}$

c $\dfrac{x^2 + 5x}{x + 5} = \dfrac{x(x + \text{......})}{x + 5}$ $= \text{..........}$

2 Simplify

a $\dfrac{x^2 - 2x - 8}{x^2 + 3x + 2} = \dfrac{(\text{..........})(\text{..........})}{(\text{..........})(\text{..........})}$ $= \text{..........}$

b $\dfrac{x^2 + 2x - 3}{x^2 + x - 2}$ ....

3 Simplify fully

a $\dfrac{x^2 - 16}{x^2 + 5x + 4}$ ....

b $\dfrac{x^2 - 10x + 25}{x^2 - 5x}$ ....

4 Simplify

a $\dfrac{2 - x}{x^2 - 6x + 8} = \dfrac{2 - x}{(\text{..........})(\text{..........})}$ $= \dfrac{-1(x - 2)}{(\text{..........})(\text{..........})}$ $= \text{..........}$

b $\dfrac{12 - 4x}{x^2 + 2x - 15}$ ....

Exam-style question

5 Simplify fully $\dfrac{2x^2 - 9x - 5}{x^2 - 4x - 5}$

.... (3 marks)

Reflect How is simplifying algebraic fractions similar to simplifying numerical fractions?

2 Adding and subtracting algebraic fractions

To add or subtract algebraic fractions:
- factorise if possible
- find the lowest common multiple (LCM) of the denominators
- write equivalent fractions with this LCM as denominator.

Guided practice

Write as a single fraction in its simplest form.

$$\frac{1}{3x+6}+\frac{1}{x+1}$$

Factorise

$$= \frac{1}{3(x+2)} + \frac{1}{x+1}$$

Find the LCM of the denominators: $3(x+2)(\square + \square)$

$$= \frac{1 \times (x+1)}{3(x+2)(x+1)} + \frac{3(x+2) \times 1}{3(x+2)(x+1)}$$

Write equivalent fractions with this LCM as the denominator.

Expand the brackets

$$= \frac{x + \dots\dots}{3(x+2)(x+1)} + \frac{3x + \dots\dots}{3(x+2)(x+1)}$$

Why?

$$\frac{x+1}{3(x+2)(x+1)} = \frac{1}{3(x+2)}$$

$$\frac{3(x+2)}{3(x+2)(x+1)} = \frac{1}{x+1}$$

Add

$$= \frac{4x+7}{3(x+2)(x+1)}$$

1 Write as a single fraction in its simplest form.

a $\frac{1}{x+1}+\frac{1}{x+3}$

Hint LCM $= (x+1)(x+3)$

$$= \frac{1 \times (\dots\dots\dots\dots)}{(x+1)(\dots\dots\dots\dots)} + \frac{1 \times (\dots\dots\dots\dots)}{(\dots\dots\dots\dots)(x+3)}$$

$= \dots\dots\dots\dots$

b $\frac{1}{x-4}+\frac{1}{x+2}$

Hint $x + 2 + (x - 4) = 2x - 2$

$= \dots\dots\dots\dots$

c $\frac{1}{x-3}-\frac{1}{x-1}$

$= \dots\dots\dots\dots$

2 Write as a single fraction.

a $\frac{2}{x+5}+\frac{1}{x+1}$

$$= \frac{2 \times (\dots\dots\dots\dots)}{(x+5)(\dots\dots\dots\dots)} + \frac{1 \times (\dots\dots\dots\dots)}{(\dots\dots\dots\dots)(x+1)}$$

$= \dots\dots\dots\dots$

b $\frac{4}{x-1}-\frac{3}{x+2}$

$= \dots\dots\dots\dots$

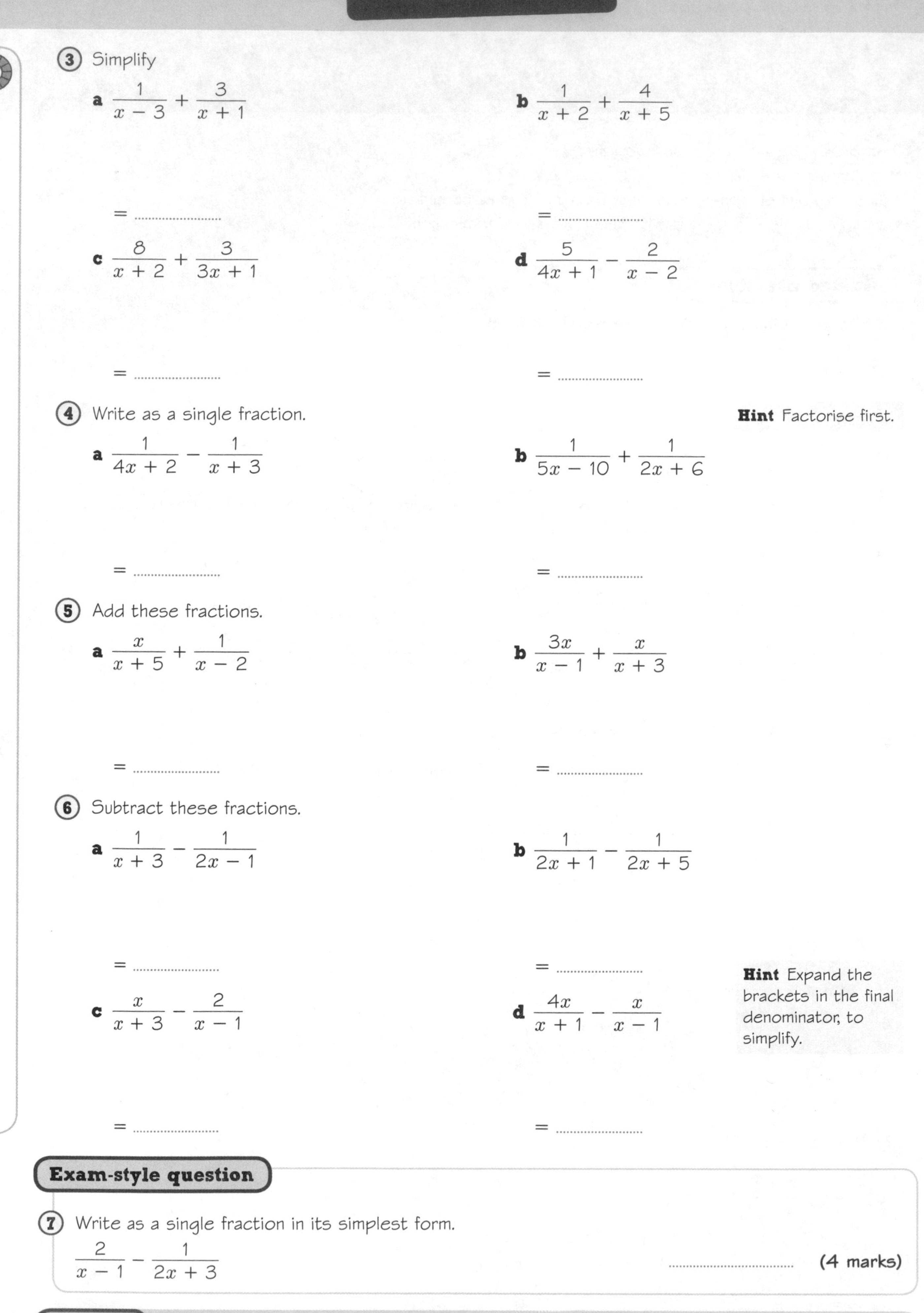

3 Simplify

a $\frac{1}{x-3} + \frac{3}{x+1}$

= ....

b $\frac{1}{x+2} + \frac{4}{x+5}$

= ....

c $\frac{8}{x+2} + \frac{3}{3x+1}$

= ....

d $\frac{5}{4x+1} - \frac{2}{x-2}$

= ....

4 Write as a single fraction.

Hint Factorise first.

a $\frac{1}{4x+2} - \frac{1}{x+3}$

= ....

b $\frac{1}{5x-10} + \frac{1}{2x+6}$

= ....

5 Add these fractions.

a $\frac{x}{x+5} + \frac{1}{x-2}$

= ....

b $\frac{3x}{x-1} + \frac{x}{x+3}$

= ....

6 Subtract these fractions.

a $\frac{1}{x+3} - \frac{1}{2x-1}$

= ....

b $\frac{1}{2x+1} - \frac{1}{2x+5}$

= ....

Hint Expand the brackets in the final denominator, to simplify.

c $\frac{x}{x+3} - \frac{2}{x-1}$

= ....

d $\frac{4x}{x+1} - \frac{x}{x-1}$

= ....

Exam-style question

7 Write as a single fraction in its simplest form.

$\frac{2}{x-1} - \frac{1}{2x+3}$

.... **(4 marks)**

Reflect How is adding algebraic fractions similar to adding numerical fractions?

3 Solving equations involving algebraic fractions

To solve equations involving algebraic fractions:

- write fraction calculations as a single fraction
- multiply both sides by the denominator
- solve the resulting equation.

Guided practice

Worked exam question

Solve $\frac{1}{x-3} + \frac{3}{x+1} = 1$

Write the LHS as a single fraction:

$\frac{4x-8}{(x-3)(x+1)} = 1$ — See skills boost 2 Q3a

Multiply both sides by the denominator.

$\frac{\cancel{(x-3)}\cancel{(x+1)}(4x-8)}{\cancel{(x-3)}\cancel{(x+1)}} = (x - \text{.............})(x + \text{.............})$ — 'Get rid' of the denominator.

Expand the brackets.

$4x - 8 = x^2 - 2x - 3$

$(x-3)(x+1)$
$= x^2 - 3x + x - 3$
$= x^2 \square - 3$

Rearrange so one side is 0.

$0 = x^2 - 6x + 5$

Solve by factorising (or use the quadratic formula).

$0 = (x - \text{.............})(x - \text{.............})$ — Factorise.

$x = 5$ or $x = 1$ — Give both solutions.

1 Solve

a $\frac{7x+3}{(x+1)(x+3)} = 1$

$7x + 3 = (\text{........................})(\text{........................})$

$7x + 3 = x^2 + \text{........................}$

$0 = x^2 - \text{........................}$

b $\frac{4x+2}{(x+2)(x-3)} = 3$

Hint Factorise and solve.

2 Solve

a $\frac{1}{x+2} + \frac{6}{x+6} = 1$

b $\frac{1}{x+2} + \frac{1}{2x-1} = 2$

Give your answer to 2 d.p.

Hint Use the quadratic formula.

Exam-style question

3 Solve $\frac{9}{x+2} - \frac{1}{3x-2} = 2$

.... **(4 marks)**

Reflect Why do these equations each have two possible solutions?

Practise the methods

Answer this question to check where to start.

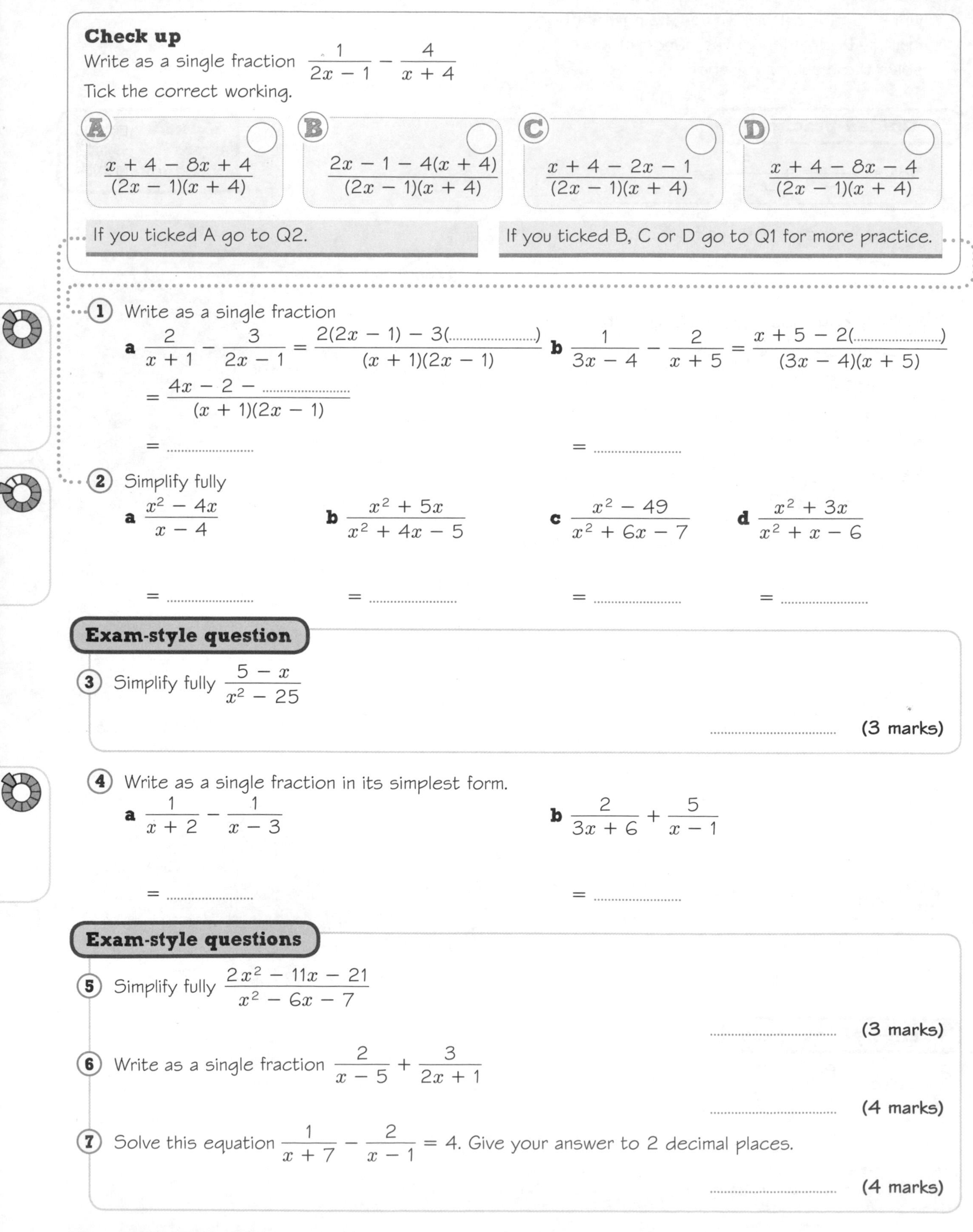

Check up

Write as a single fraction $\frac{1}{2x - 1} - \frac{4}{x + 4}$

Tick the correct working.

A $\frac{x + 4 - 8x + 4}{(2x - 1)(x + 4)}$

B $\frac{2x - 1 - 4(x + 4)}{(2x - 1)(x + 4)}$

C $\frac{x + 4 - 2x - 1}{(2x - 1)(x + 4)}$

D $\frac{x + 4 - 8x - 4}{(2x - 1)(x + 4)}$

If you ticked A go to Q2.

If you ticked B, C or D go to Q1 for more practice.

1 Write as a single fraction

a $\frac{2}{x + 1} - \frac{3}{2x - 1} = \frac{2(2x - 1) - 3(\ldots\ldots)}{(x + 1)(2x - 1)}$

$= \frac{4x - 2 - \ldots\ldots}{(x + 1)(2x - 1)}$

$= \ldots\ldots$

b $\frac{1}{3x - 4} - \frac{2}{x + 5} = \frac{x + 5 - 2(\ldots\ldots)}{(3x - 4)(x + 5)}$

$= \ldots\ldots$

2 Simplify fully

a $\frac{x^2 - 4x}{x - 4}$ $= \ldots\ldots$

b $\frac{x^2 + 5x}{x^2 + 4x - 5}$ $= \ldots\ldots$

c $\frac{x^2 - 49}{x^2 + 6x - 7}$ $= \ldots\ldots$

d $\frac{x^2 + 3x}{x^2 + x - 6}$ $= \ldots\ldots$

Exam-style question

3 Simplify fully $\frac{5 - x}{x^2 - 25}$

.... **(3 marks)**

4 Write as a single fraction in its simplest form.

a $\frac{1}{x + 2} - \frac{1}{x - 3}$ $= \ldots\ldots$

b $\frac{2}{3x + 6} + \frac{5}{x - 1}$ $= \ldots\ldots$

Exam-style questions

5 Simplify fully $\frac{2x^2 - 11x - 21}{x^2 - 6x - 7}$

.... **(3 marks)**

6 Write as a single fraction $\frac{2}{x - 5} + \frac{3}{2x + 1}$

.... **(4 marks)**

7 Solve this equation $\frac{1}{x + 7} - \frac{2}{x - 1} = 4$. Give your answer to 2 decimal places.

.... **(4 marks)**

Problem-solve!

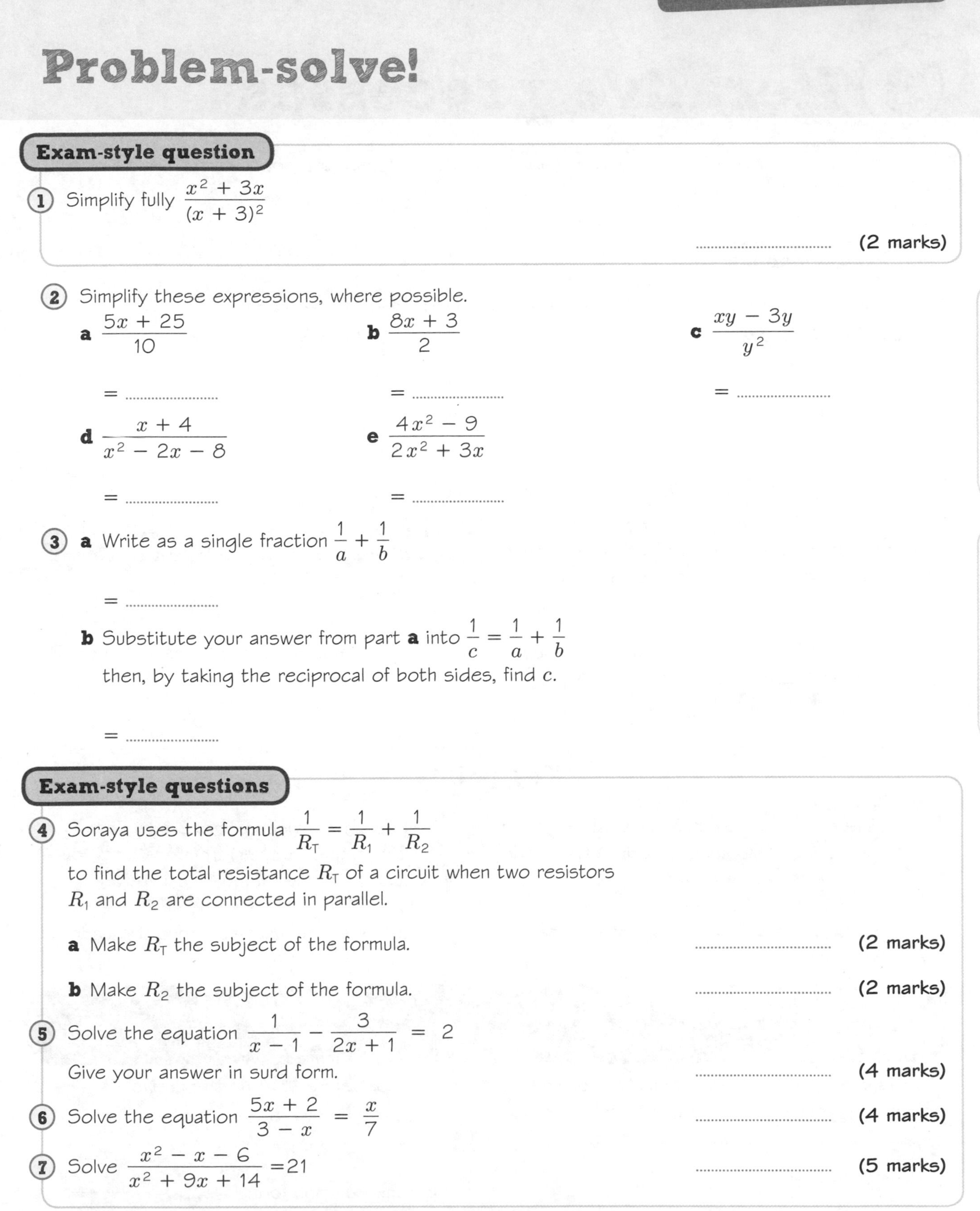

Exam-style question

1. Simplify fully $\frac{x^2 + 3x}{(x + 3)^2}$ .... **(2 marks)**

2. Simplify these expressions, where possible.

a $\frac{5x + 25}{10}$ $= $ ....

b $\frac{8x + 3}{2}$ $= $ ....

c $\frac{xy - 3y}{y^2}$ $= $ ....

d $\frac{x + 4}{x^2 - 2x - 8}$ $= $ ....

e $\frac{4x^2 - 9}{2x^2 + 3x}$ $= $ ....

3. **a** Write as a single fraction $\frac{1}{a} + \frac{1}{b}$

$= $ ....

b Substitute your answer from part **a** into $\frac{1}{c} = \frac{1}{a} + \frac{1}{b}$ then, by taking the reciprocal of both sides, find c.

$= $ ....

Exam-style questions

4. Soraya uses the formula $\frac{1}{R_T} = \frac{1}{R_1} + \frac{1}{R_2}$ to find the total resistance R_T of a circuit when two resistors R_1 and R_2 are connected in parallel.

a Make R_T the subject of the formula. .... **(2 marks)**

b Make R_2 the subject of the formula. .... **(2 marks)**

5. Solve the equation $\frac{1}{x - 1} - \frac{3}{2x + 1} = 2$
Give your answer in surd form. .... **(4 marks)**

6. Solve the equation $\frac{5x + 2}{3 - x} = \frac{x}{7}$ .... **(4 marks)**

7. Solve $\frac{x^2 - x - 6}{x^2 + 9x + 14} = 21$ .... **(5 marks)**

Now that you have completed this unit, how confident do you feel?

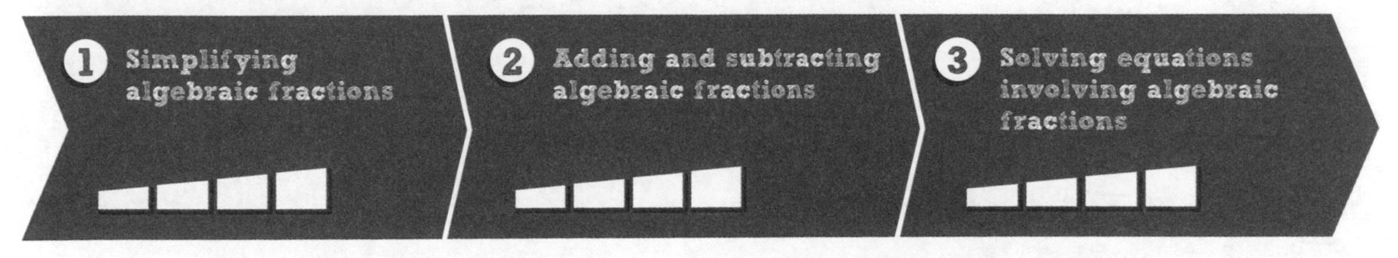

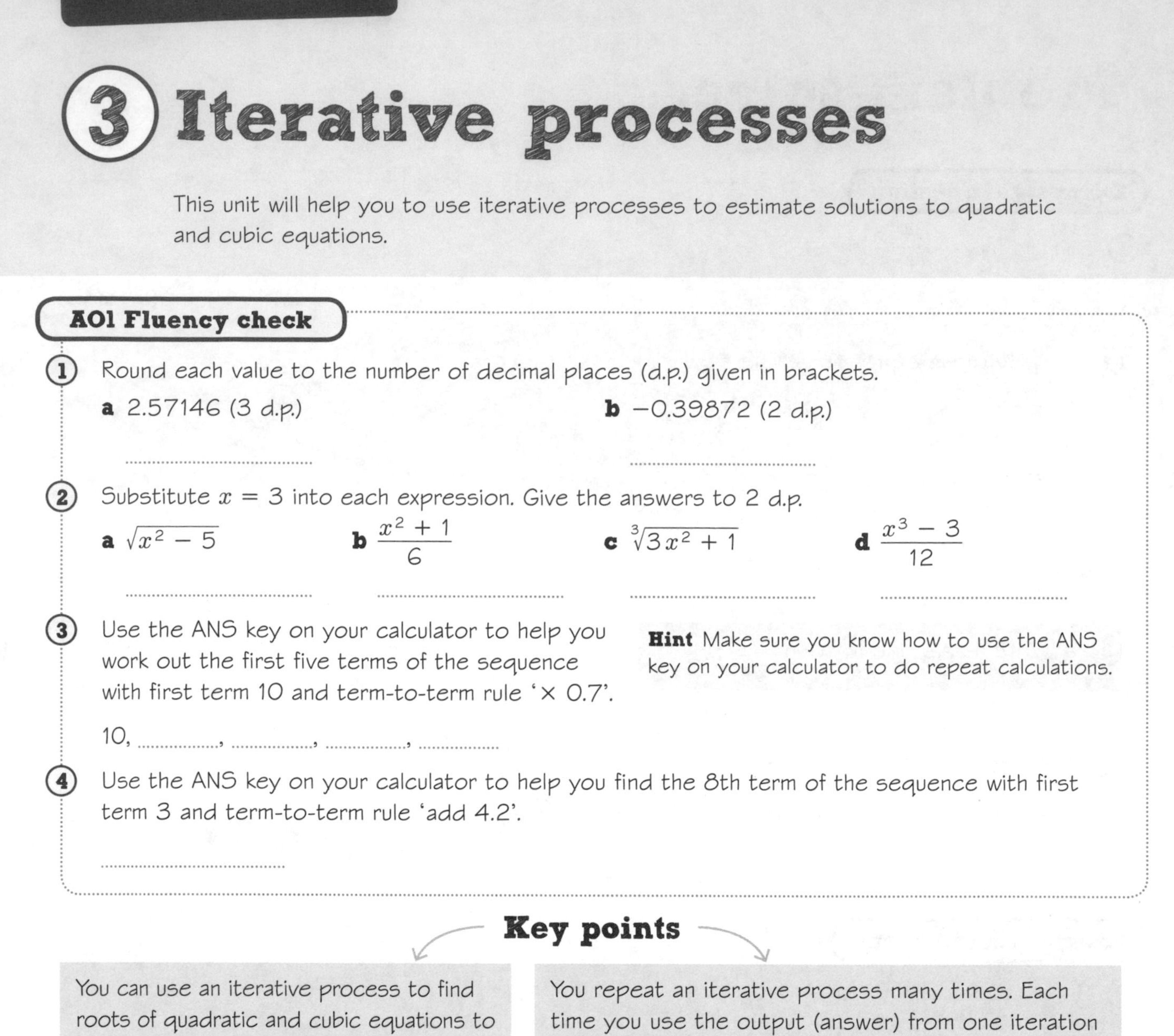

3 Iterative processes

This unit will help you to use iterative processes to estimate solutions to quadratic and cubic equations.

AO1 Fluency check

1. Round each value to the number of decimal places (d.p.) given in brackets.

   a 2.57146 (3 d.p.) ....

   b −0.39872 (2 d.p.) ....

2. Substitute $x = 3$ into each expression. Give the answers to 2 d.p.

   a $\sqrt{x^2 - 5}$ ....

   b $\dfrac{x^2 + 1}{6}$ ....

   c $\sqrt[3]{3x^2 + 1}$ ....

   d $\dfrac{x^3 - 3}{12}$ ....

3. Use the ANS key on your calculator to help you work out the first five terms of the sequence with first term 10 and term-to-term rule '× 0.7'.

   Hint Make sure you know how to use the ANS key on your calculator to do repeat calculations.

   10, ..., ..., ..., ...

4. Use the ANS key on your calculator to help you find the 8th term of the sequence with first term 3 and term-to-term rule 'add 4.2'.

   ....

Key points

You can use an iterative process to find roots of quadratic and cubic equations to a given number of decimal places (d.p.).

You repeat an iterative process many times. Each time you use the output (answer) from one iteration as the input into the formula for the next iteration.

These **skills boosts** will help you to use iterative processes to estimate solutions to quadratic and cubic equations.

1 Solving quadratic equations using an iterative process

2 Solving cubic equations using an iterative process

You might have already done some work on iterative processes. Before starting the first skills boost, rate your confidence using each concept.

1 Use the iteration formula $x_{n+1} = \sqrt{3 - x_n}$ with $x_0 = -1$ to estimate a solution to $x^2 + x - 3 = 0$ to 2 d.p.

....

....

2 Use the iteration formula $x_{n+1} = \dfrac{x_n^{\,3} - 7}{8}$ with $x_0 = 1$ to estimate a solution to $x^3 - 8x - 7 = 0$ to 2 d.p.

....

....

How confident are you?

1 Solving quadratic equations using an iterative process

Use the ANS key on your calculator to input the previous answer into the iteration formula.

Guided practice

Use the iteration formula $x_{n+1} = \sqrt{2x_n + 5}$
to find an estimated solution to $x^2 - 2x - 5 = 0$, to 1 d.p.
Start with $x_0 = 3$

Write down the iteration formula.

$x_{n+1} = \sqrt{2x_n + 5}$

Substitute $x_0 = 3$ for x_n in the formula.

Work out x_1.

$x_1 = \sqrt{2x_0 + 5}$

$x_1 = \sqrt{2 \times 3 + 5}$

$x_1 = \sqrt{11} = 3.316\,624\,79$

Substitute x_1 to work out x_2.

Use ANS on your calculator.

$x_2 = \sqrt{2 \times 3.316\,624\,79 + 5}$

$x_2 = \sqrt{(2 \times \text{ANS} + 5)}$

$x_2 = 3.410\,754\,987 = 3.4$ (to 1 d.p.)

Is the value the same as the previous value, to 1 d.p.?

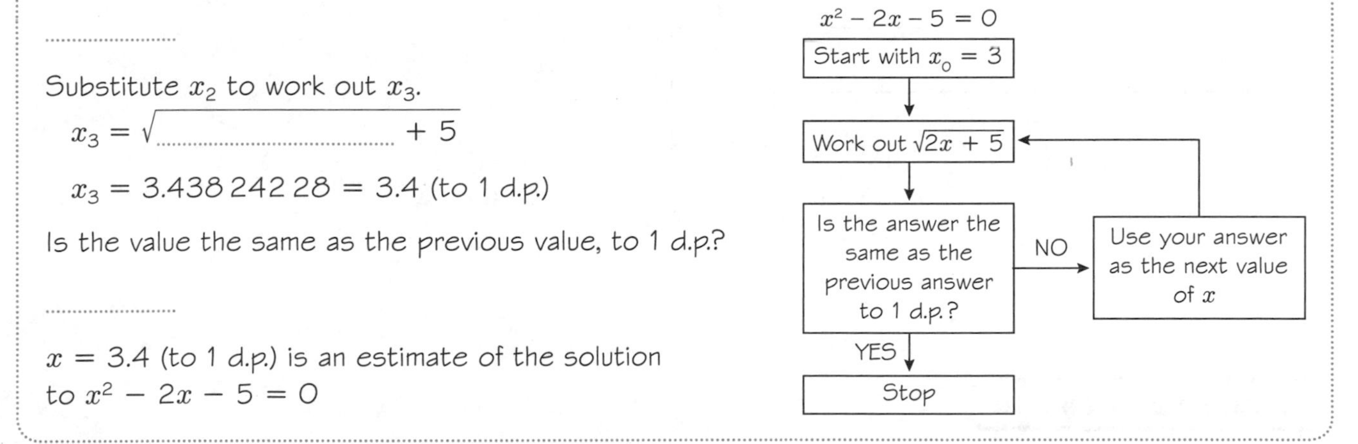

....

Substitute x_2 to work out x_3.

$x_3 = \sqrt{\text{..........................} + 5}$

$x_3 = 3.438\,242\,28 = 3.4$ (to 1 d.p.)

Is the value the same as the previous value, to 1 d.p.?

....

$x = 3.4$ (to 1 d.p.) is an estimate of the solution to $x^2 - 2x - 5 = 0$

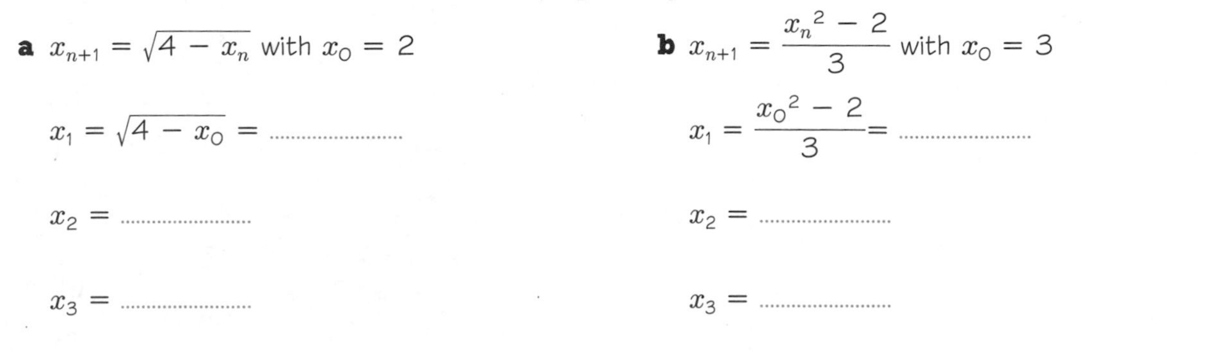

1 Use each iteration formula to calculate x_1, x_2 and x_3.

a $x_{n+1} = \sqrt{4 - x_n}$ with $x_0 = 2$

$x_1 = \sqrt{4 - x_0} =$ ....

$x_2 =$ ....

$x_3 =$ ....

b $x_{n+1} = \dfrac{x_n{}^2 - 2}{3}$ with $x_0 = 3$

$x_1 = \dfrac{x_0{}^2 - 2}{3} =$ ....

$x_2 =$ ....

$x_3 =$ ....

② Use the iteration formula

$x_{n+1} = \sqrt{x_n + 7}$

to find an estimated solution to $x^2 - x - 7 = 0$, to 2 d.p.

Start with $x_0 = 3$.

$x_{n+1} = \sqrt{x_n + 7}$

$x_1 = \sqrt{x_0 + 7}$

$x_1 = \sqrt{\ldots\ldots + 7}$

$x_1 = \ldots\ldots\ldots\ldots = \ldots\ldots$ (to 2 d.p.)

$x_2 = \sqrt{x_1 + 7} = \sqrt{\ldots\ldots + 7}$

$x_2 = \ldots\ldots\ldots\ldots = \ldots\ldots$ (to 2 d.p.)

$x_3 = \ldots\ldots\ldots\ldots\ldots\ldots$

$x_3 = \ldots\ldots\ldots\ldots = \ldots\ldots$ (to 2 d.p.)

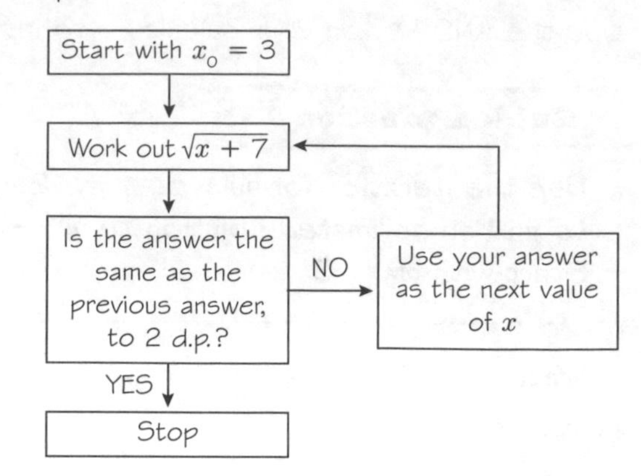

③ Use the iteration formula

$$x_{n+1} = \frac{x_n^2 - 3}{8}$$

to find an estimated solution to $x^2 - 8x - 3 = 0$, to 3 d.p.

Start with $x_0 = -1$

Hint For an estimate to 3 d.p., round your answers to 3 d.p.

	To 3 d.p.	Same as previous answer?
x_0	−1.000	
x_1	−0.250	No
x_2		
x_3		

Exam-style question

④ Use the iteration formula

$x_{n+1} = \sqrt{32 - 2x_n}$

to find an estimated solution to $x^2 + 2x - 32 = 0$.
Start with $x_0 = 4.5$
Give your estimate to 2 d.p.

.... **(4 marks)**

Reflect Does using a table like the one in Q3 help you to organise and show your working?

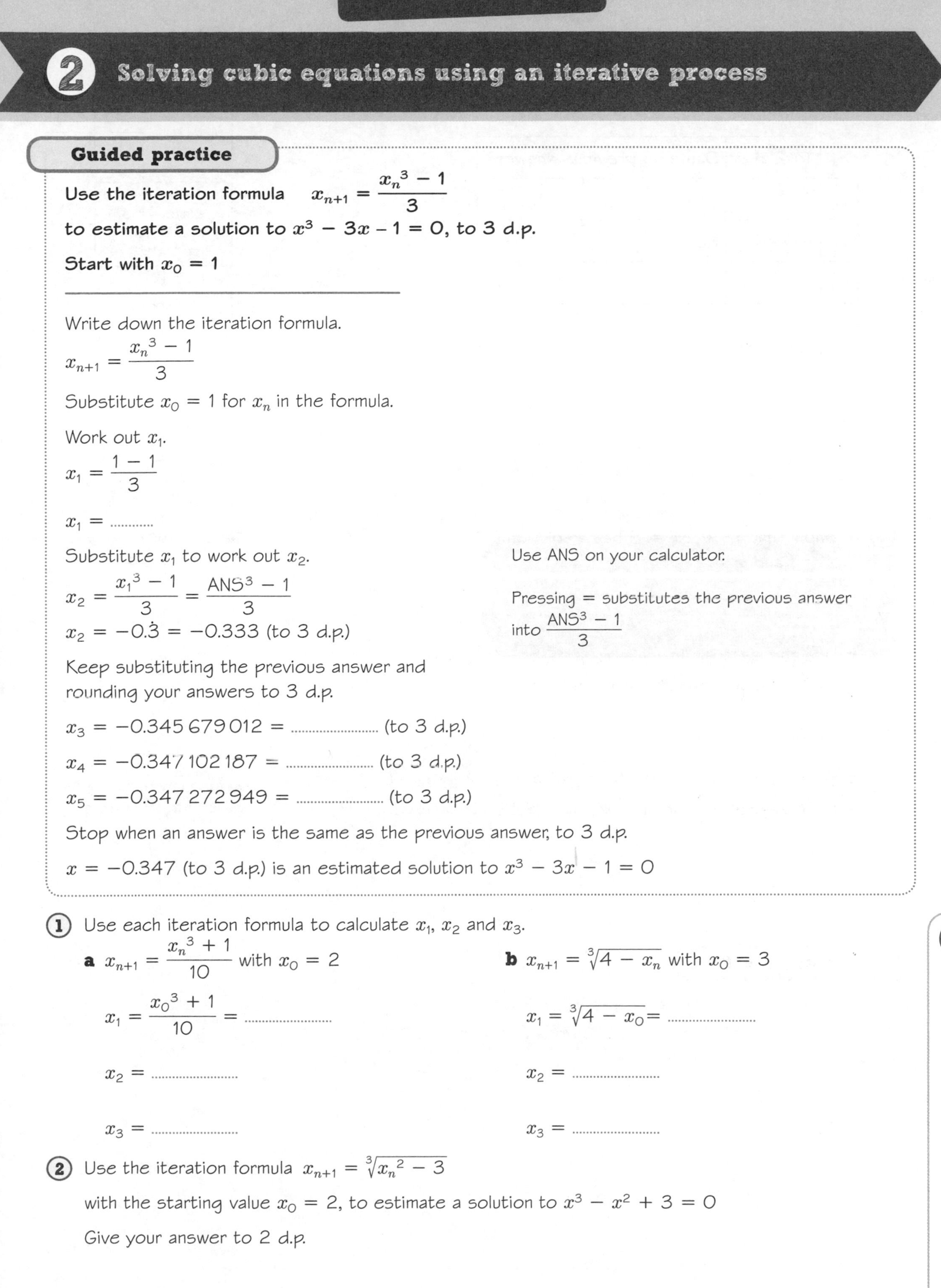

2 Solving cubic equations using an iterative process

Guided practice

Use the iteration formula $x_{n+1} = \dfrac{x_n^3 - 1}{3}$

to estimate a solution to $x^3 - 3x - 1 = 0$, to 3 d.p.

Start with $x_0 = 1$

Write down the iteration formula.

$x_{n+1} = \dfrac{x_n^3 - 1}{3}$

Substitute $x_0 = 1$ for x_n in the formula.

Work out x_1.

$x_1 = \dfrac{1 - 1}{3}$

$x_1 =$ ....

Substitute x_1 to work out x_2.

$x_2 = \dfrac{x_1^3 - 1}{3} = \dfrac{\text{ANS}^3 - 1}{3}$

$x_2 = -0.\dot{3} = -0.333$ (to 3 d.p.)

Use ANS on your calculator.

Pressing = substitutes the previous answer into $\dfrac{\text{ANS}^3 - 1}{3}$

Keep substituting the previous answer and rounding your answers to 3 d.p.

$x_3 = -0.345\,679\,012 =$ .... (to 3 d.p.)

$x_4 = -0.347\,102\,187 =$ .... (to 3 d.p.)

$x_5 = -0.347\,272\,949 =$ .... (to 3 d.p.)

Stop when an answer is the same as the previous answer, to 3 d.p.

$x = -0.347$ (to 3 d.p.) is an estimated solution to $x^3 - 3x - 1 = 0$

1 Use each iteration formula to calculate x_1, x_2 and x_3.

a $x_{n+1} = \dfrac{x_n^3 + 1}{10}$ with $x_0 = 2$

$x_1 = \dfrac{x_0^3 + 1}{10} =$ ....

$x_2 =$ ....

$x_3 =$ ....

b $x_{n+1} = \sqrt[3]{4 - x_n}$ with $x_0 = 3$

$x_1 = \sqrt[3]{4 - x_0} =$ ....

$x_2 =$ ....

$x_3 =$ ....

2 Use the iteration formula $x_{n+1} = \sqrt[3]{x_n^2 - 3}$

with the starting value $x_0 = 2$, to estimate a solution to $x^3 - x^2 + 3 = 0$

Give your answer to 2 d.p.

....

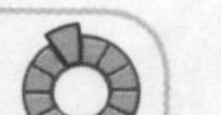

③ Use the iteration formula $x_{n+1} = \frac{x_n^3 - 1}{4}$ to find an estimated solution to $x^3 - 4x - 1 = 0$ to 2 d.p. Start with $x_0 = 0$

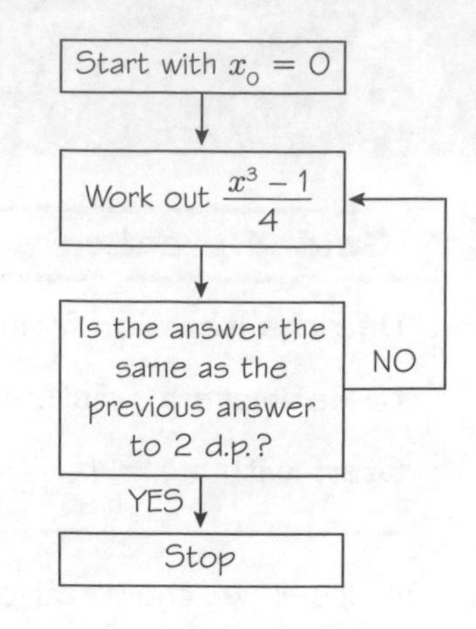

	To 2 d.p.	Same as previous answer?
x_0	0	
x_1	−0.25	No
x_2		
x_3		
x_4		

....

④ **a** Show that $x^3 + 2x = 5$ has a solution between $x = 1$ and $x = 2$

Hint Substitute $x = 1$ and $x = 2$ into $x^3 + 2x$

....

b Use the iteration formula $x_{n+1} = \sqrt[3]{5 - 2x_n}$, with $x_0 = 1.5$ to find an approximate solution to $x^3 + 2x = 5$, to 2 d.p.

....

c Substitute your solution from part **a** into $x^3 + 2x$ to show that it is an approximate solution to $x^3 + 2x = 5$

Hint Show that for your value of x, $x^3 + 2x \approx 5$

....

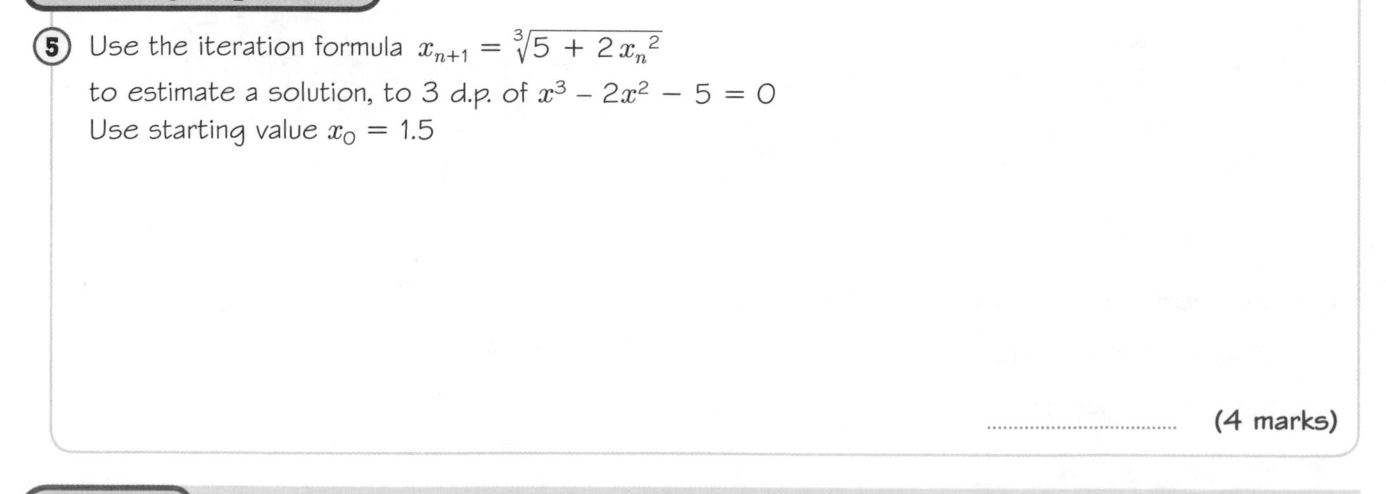

Exam-style question

⑤ Use the iteration formula $x_{n+1} = \sqrt[3]{5 + 2x_n^2}$ to estimate a solution, to 3 d.p. of $x^3 - 2x^2 - 5 = 0$
Use starting value $x_0 = 1.5$

.... **(4 marks)**

Reflect How is using an iteration formula to estimate solutions to cubic formulae similar to using an iteration formula to estimate solutions to quadratic formulae?

Practise the methods

Answer this question to check where to start.

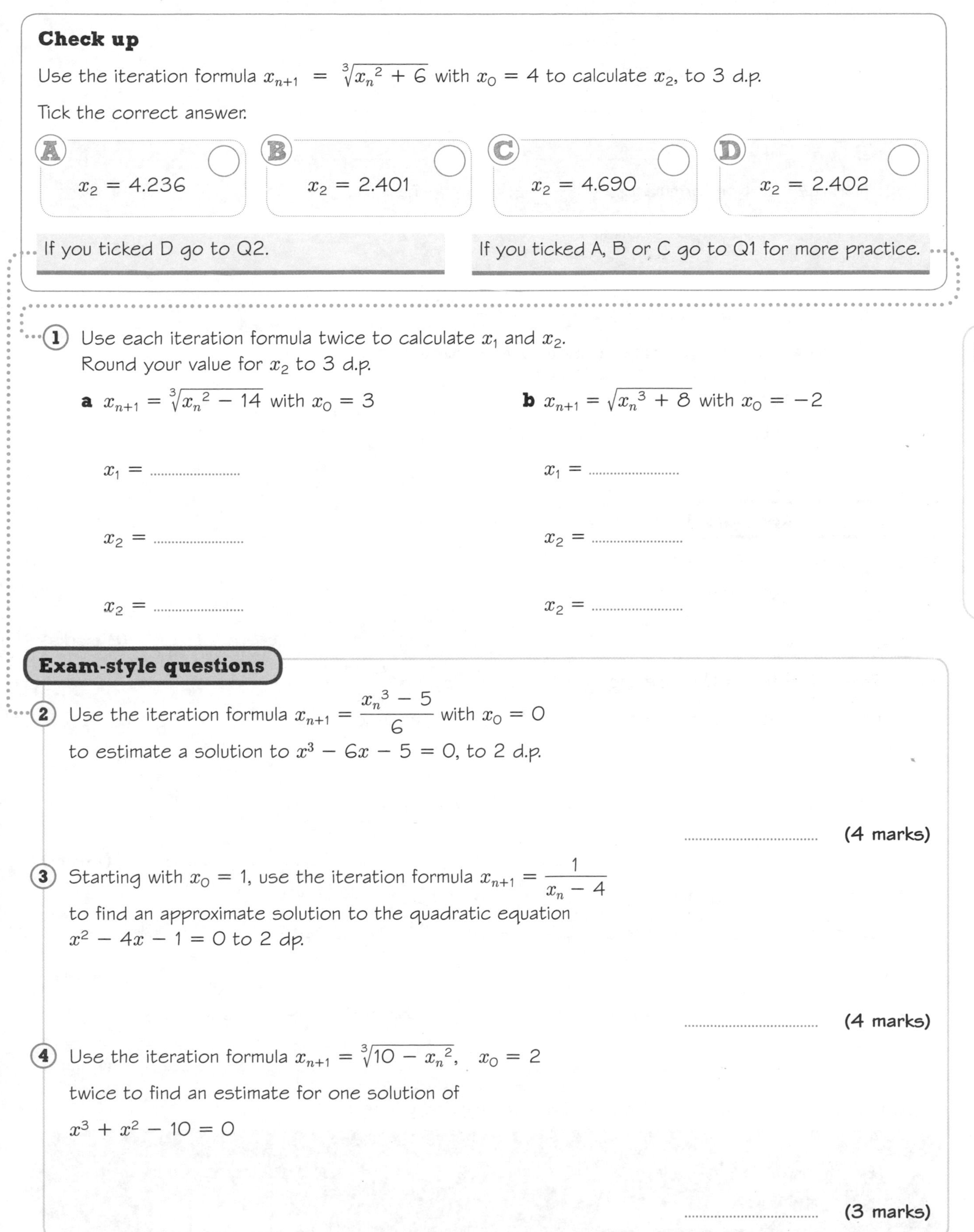

Check up

Use the iteration formula $x_{n+1} = \sqrt[3]{x_n^2 + 6}$ with $x_0 = 4$ to calculate x_2, to 3 d.p.

Tick the correct answer.

A $x_2 = 4.236$

B $x_2 = 2.401$

C $x_2 = 4.690$

D $x_2 = 2.402$

If you ticked D go to Q2.

If you ticked A, B or C go to Q1 for more practice.

1 Use each iteration formula twice to calculate x_1 and x_2.
Round your value for x_2 to 3 d.p.

a $x_{n+1} = \sqrt[3]{x_n^2 - 14}$ with $x_0 = 3$

$x_1 =$ ....

$x_2 =$ ....

$x_2 =$ ....

b $x_{n+1} = \sqrt{x_n^3 + 8}$ with $x_0 = -2$

$x_1 =$ ....

$x_2 =$ ....

$x_2 =$ ....

Exam-style questions

2 Use the iteration formula $x_{n+1} = \dfrac{x_n^3 - 5}{6}$ with $x_0 = 0$
to estimate a solution to $x^3 - 6x - 5 = 0$, to 2 d.p.

.... **(4 marks)**

3 Starting with $x_0 = 1$, use the iteration formula $x_{n+1} = \dfrac{1}{x_n - 4}$
to find an approximate solution to the quadratic equation
$x^2 - 4x - 1 = 0$ to 2 dp.

.... **(4 marks)**

4 Use the iteration formula $x_{n+1} = \sqrt[3]{10 - x_n^2}$, $x_0 = 2$
twice to find an estimate for one solution of
$x^3 + x^2 - 10 = 0$

.... **(3 marks)**

Problem-solve!

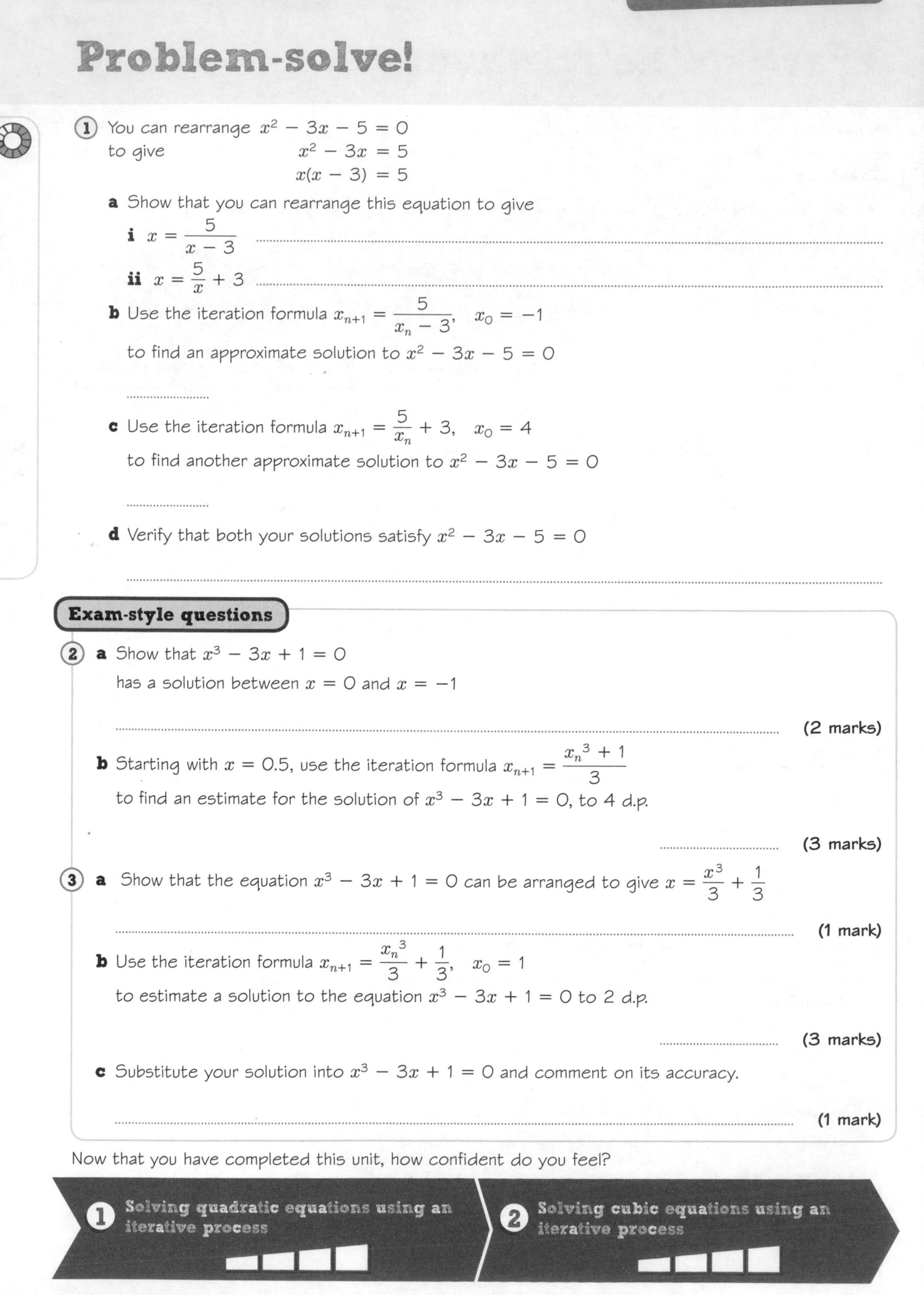

1. You can rearrange $x^2 - 3x - 5 = 0$
   to give $x^2 - 3x = 5$
   $x(x - 3) = 5$

   a Show that you can rearrange this equation to give

   i $x = \dfrac{5}{x - 3}$ ....

   ii $x = \dfrac{5}{x} + 3$ ....

   b Use the iteration formula $x_{n+1} = \dfrac{5}{x_n - 3}$, $x_0 = -1$

   to find an approximate solution to $x^2 - 3x - 5 = 0$

   ....

   c Use the iteration formula $x_{n+1} = \dfrac{5}{x_n} + 3$, $x_0 = 4$

   to find another approximate solution to $x^2 - 3x - 5 = 0$

   ....

   d Verify that both your solutions satisfy $x^2 - 3x - 5 = 0$

   ....

Exam-style questions

2. **a** Show that $x^3 - 3x + 1 = 0$
   has a solution between $x = 0$ and $x = -1$

   .... **(2 marks)**

   b Starting with $x = 0.5$, use the iteration formula $x_{n+1} = \dfrac{{x_n}^3 + 1}{3}$

   to find an estimate for the solution of $x^3 - 3x + 1 = 0$, to 4 d.p.

   .... **(3 marks)**

3. **a** Show that the equation $x^3 - 3x + 1 = 0$ can be arranged to give $x = \dfrac{x^3}{3} + \dfrac{1}{3}$

   .... **(1 mark)**

   b Use the iteration formula $x_{n+1} = \dfrac{{x_n}^3}{3} + \dfrac{1}{3}$, $x_0 = 1$

   to estimate a solution to the equation $x^3 - 3x + 1 = 0$ to 2 d.p.

   .... **(3 marks)**

   c Substitute your solution into $x^3 - 3x + 1 = 0$ and comment on its accuracy.

   .... **(1 mark)**

Now that you have completed this unit, how confident do you feel?

1 Solving quadratic equations using an iterative process

2 Solving cubic equations using an iterative process

4 Histograms with unequal class widths

This unit will help you to draw and interpret histograms.

AO1 Fluency check

1. Calculate the area of each rectangle.

a y: 0, 0.5, 1; x: 0, 2 ....

b y: 0, 1, 2; x: 10, 15 ....

2. The frequency table shows the ages of children in a nursery.

Age, a (months)	Frequency
$0 \leqslant a < 8$	4
$8 \leqslant a < 12$	6
$12 \leqslant a < 15$	9
$15 \leqslant a < 18$	6
$18 \leqslant a < 24$	3

a Find the modal class. ....

b Estimate the range. ....

c Find the class interval containing the median. ....

d Calculate an estimate for the mean age, to the nearest month. ....

Key points

In a histogram, the **area** of the bar represents the frequency.

The vertical axis is the **frequency density**.

These **skills boosts** will help you to draw and interpret histograms.

1 Drawing histograms
2 Interpreting histograms
3 Estimating averages and range from a histogram

You might have already done some work on histograms. Before starting the first skills boost, rate your confidence with these questions.

1. Draw a histogram for the data in Q2 above.

2. The histogram shows the lengths of earthworms.

Lengths of earthworms

Frequency density: 0, 0.5, 1, 1.5
Length (cm): 0, 5, 10, 15, 20, 25, 30

a How many earthworms were between 15 and 20 cm long? ....

b Calculate an estimate for the mean length.

....

How confident are you?

Skills boost

1 Drawing histograms

The **area** of the bar represents frequency, so

frequency density × class width = frequency

Rearranging:

$$\text{frequency density} = \frac{\text{frequency}}{\text{class width}}$$

Frequency density

class width

Guided practice

The frequency table shows the times some people took to do a puzzle.

Time, t (seconds)	Frequency
$0 \leqslant t < 20$	2
$20 \leqslant t < 25$	7
$25 \leqslant t < 35$	12
$35 \leqslant t < 40$	1

Draw a histogram to represent this data.

Add columns to the table for class width and frequency density.

Time, t (seconds)	Frequency	Class width	Frequency density
$0 \leqslant t < 20$	2	$20 - 0 = 20$	$\frac{2}{20} = 0.1$
$20 \leqslant t < 25$	7	5	$\frac{7}{5} =$ ....
$25 \leqslant t < 35$	12	....	$\frac{12}{10} = 1.2$
$35 \leqslant t < 40$	1	....	$\frac{1}{5} =$ ....

Class width = upper class boundary − lower class boundary

Draw axes for frequency density and time.

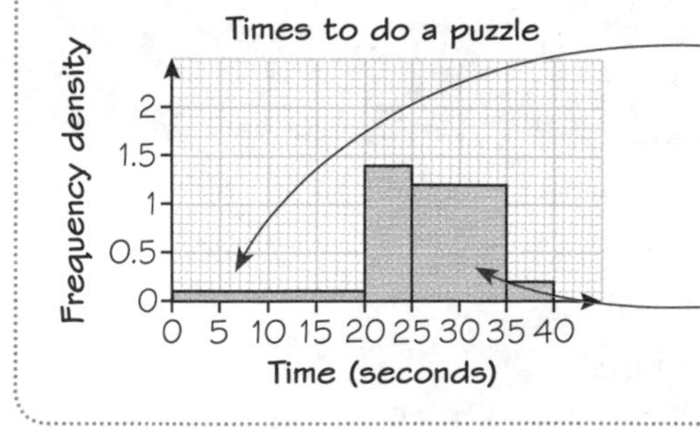

The first bar is from 0–20 on the Time axis, and height 0.1 on the Frequency density axis.

Check: Area = 10 × 1.2 = 12 = frequency ✓

① The table shows the lengths of some bees.

Length, l (mm)	Frequency	Class width	Frequency density
$5 \leqslant l < 10$	3		
$10 \leqslant l < 12$	10		
$12 \leqslant l < 15$	9		
$15 \leqslant l < 20$	17		

a Complete the columns for class width and frequency density.

b Label the frequency density axis with a suitable scale.

c Plot a histogram to show the data.

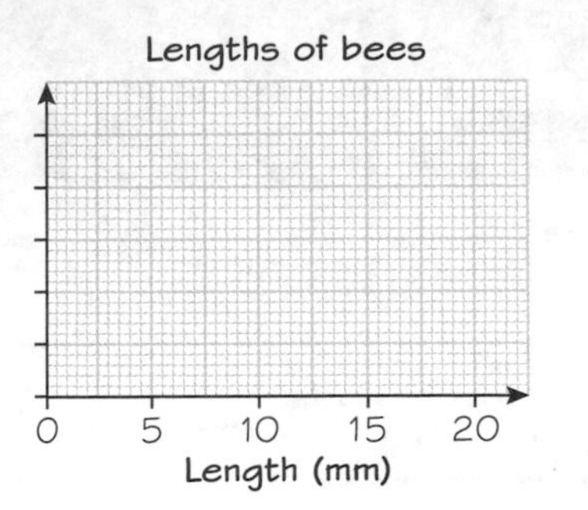

2 The table shows the lengths of 20 snakes.

Length l (m)	Frequency
$0.2 \leqslant l < 0.7$	5
$0.7 \leqslant l < 0.9$	4
$0.9 \leqslant l < 1.0$	3
$1.0 \leqslant l < 1.4$	8

Hint Add columns to the table for class width and frequency density.

Draw a histogram to represent this data.

Hint Give your histogram a title.

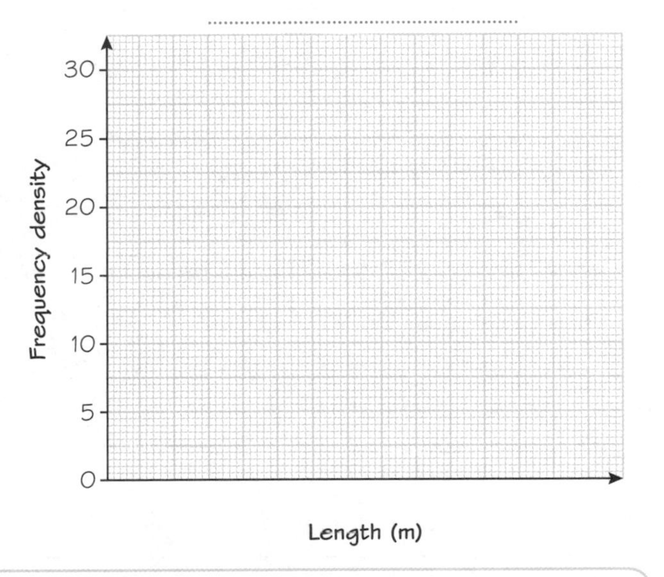

Exam-style question

3 The table shows the weights of parcels delivered on one day.

Weight, (kg)	Frequency
$0.5 \leqslant w < 1.5$	3
$1.5 \leqslant w < 3$	6
$3 \leqslant w < 5$	7
$5 \leqslant w < 10$	4

Draw a histogram to represent this data.

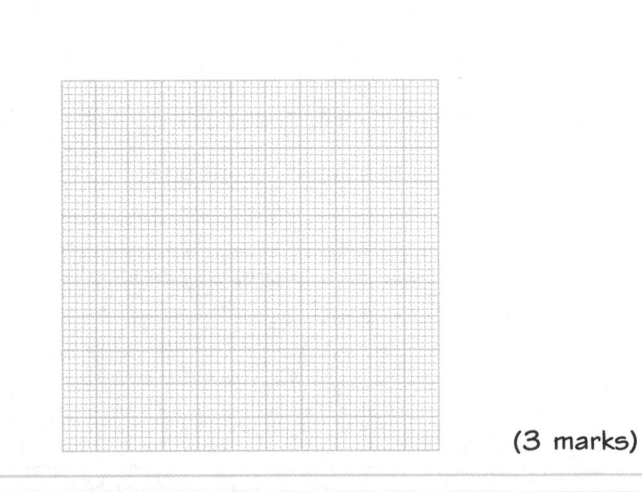

(3 marks)

Reflect Why might the class with the tallest bar not be the modal class?

2 Interpreting histograms

To find the frequency or number of items represented by one bar of a histogram, calculate its area.

Frequency = class width × frequency density

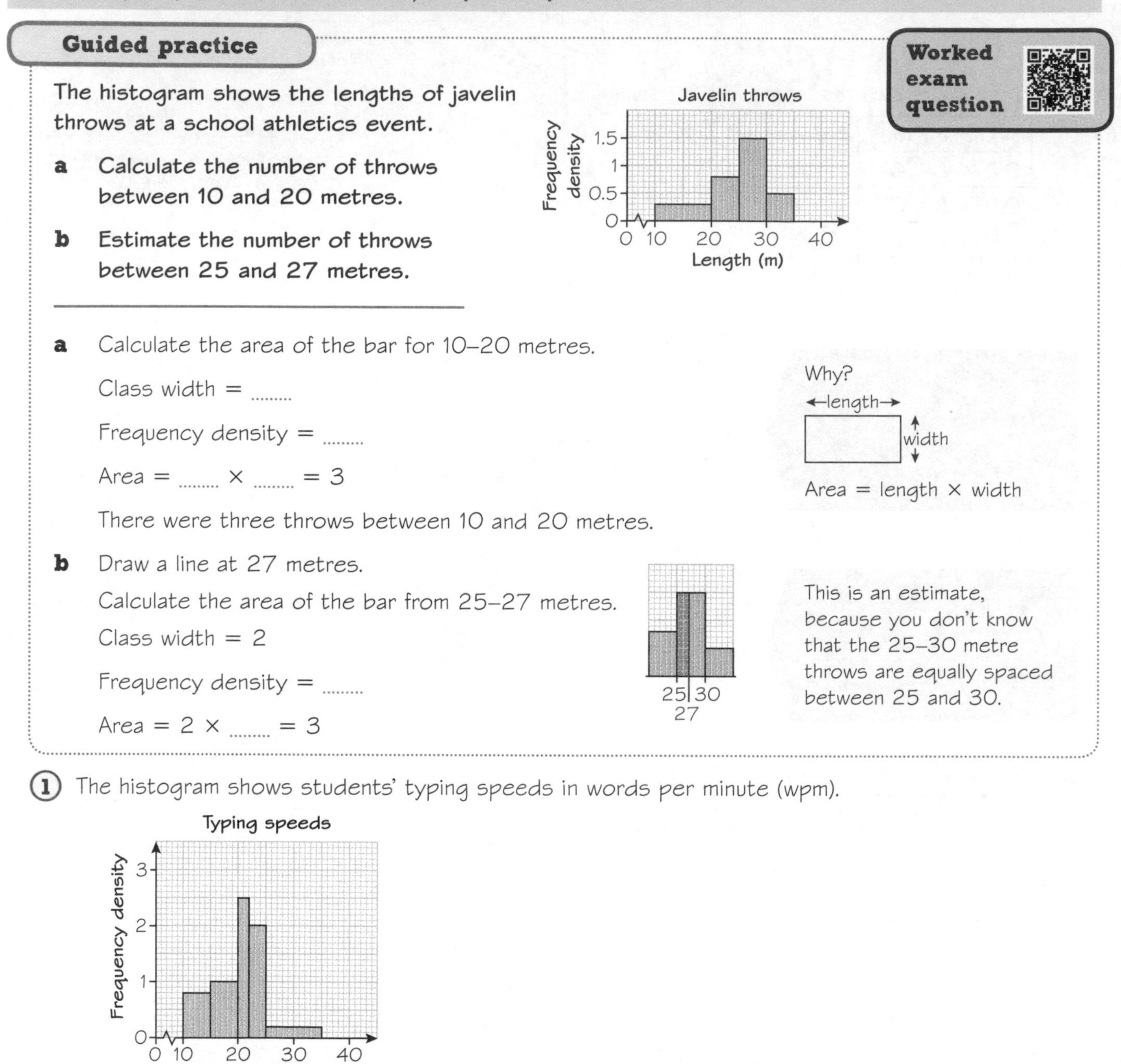

Guided practice

Worked exam question

The histogram shows the lengths of javelin throws at a school athletics event.

a **Calculate the number of throws between 10 and 20 metres.**

b **Estimate the number of throws between 25 and 27 metres.**

---

a Calculate the area of the bar for 10–20 metres.

Class width = ...

Frequency density = ...

Area = ... × .... = 3

There were three throws between 10 and 20 metres.

Why?

Area = length × width

b Draw a line at 27 metres.

Calculate the area of the bar from 25–27 metres.

Class width = 2

Frequency density = ...

Area = 2 × .... = 3

This is an estimate, because you don't know that the 25–30 metre throws are equally spaced between 25 and 30.

1. The histogram shows students' typing speeds in words per minute (wpm).

a Calculate the number of people each bar represents.

Write the values in this frequency table.

Typing speed, t (wpm)	Number of people
$10 \leqslant t < 15$	
$15 \leqslant t < 20$	
$20 \leqslant t < 22$	
$22 \leqslant t < 25$	
$25 \leqslant t < 35$	

Hint Class width = 5
Frequency density = 0.8
5 × 0.8 = □

b How many people does the histogram represent in total? ....

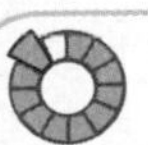

2) The histogram shows the shoulder heights of roe deer in a herd.

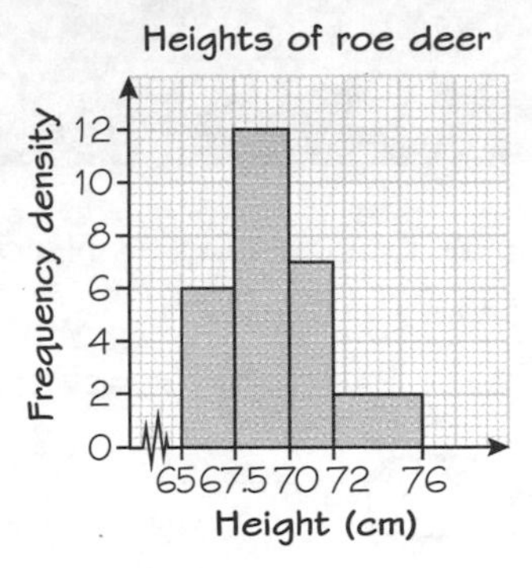

Hint Read the scales carefully.

a Complete the frequency table for this data.

Height, h (cm)	Frequency
$65 \leqslant h < 67.5$	
$67.5 \leqslant h < 70$	
$70 \leqslant h < 72$	
$72 \leqslant h < 76$	

b How many roe deer were measured? ....

3) The histogram shows the ages of visitors to a museum.

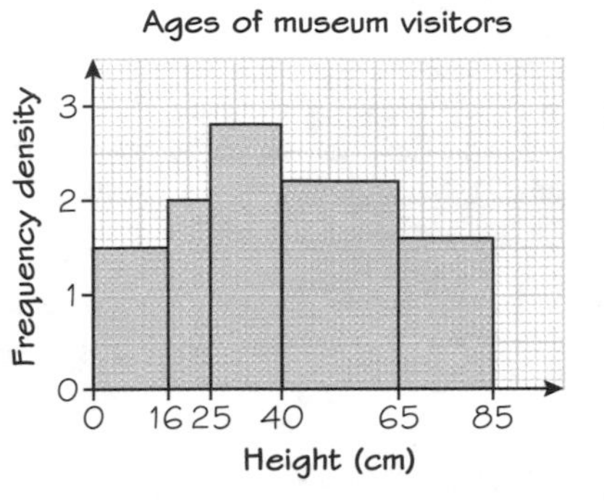

Hint Calculate the frequencies for $0 \leqslant a < 16$ and $16 \leqslant a < 25$.

a Calculate the number of under-25s. ....

b Estimate the number of people between 25 and 35. ....

Hint Draw a line at 35.

c Estimate the number of people between 70 and 80. ....

4) From the data in Q2, estimate the number of roe deer with shoulder heights

a between 72 and 75 cm ....

b between 70 and 75 cm ....

c between 67.5 cm and 75 cm. ....

Exam-style question

5) The histogram shows students' times, in minutes, to complete a maths test.

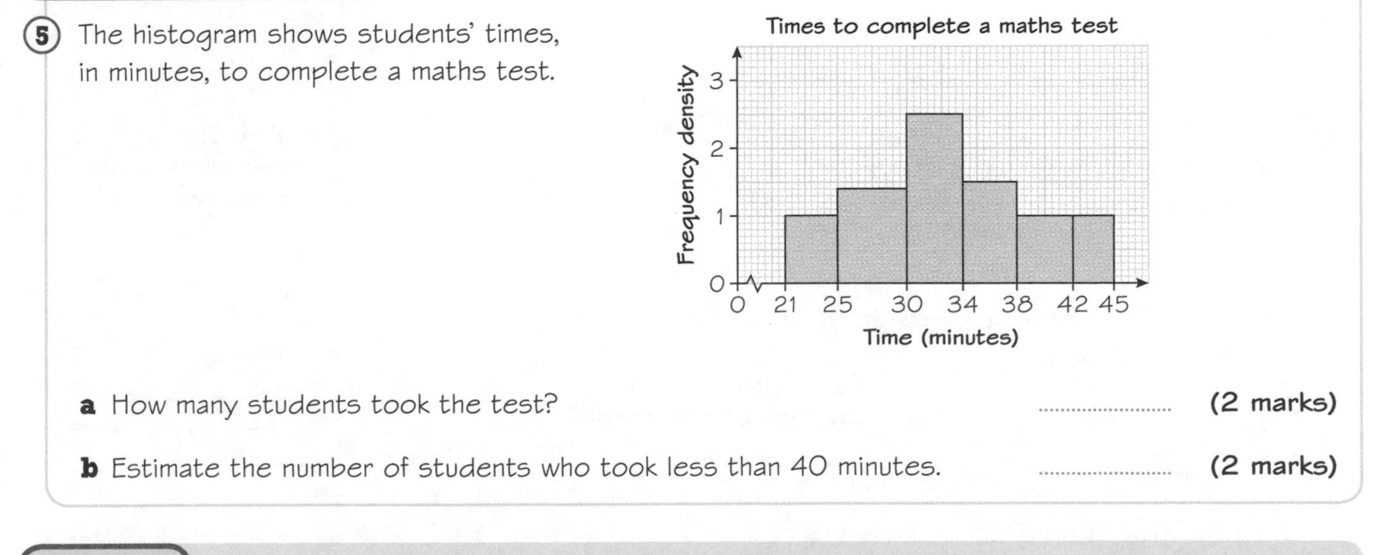

a How many students took the test? .... **(2 marks)**

b Estimate the number of students who took less than 40 minutes. .... **(2 marks)**

Reflect In Q5, the bars for $38 \leqslant t < 42$ and $42 \leqslant t < 45$ are the same height. Explain why they do not represent the same frequencies.

Estimating averages and range from a histogram

To estimate the mean and median from a histogram
- make a frequency table for the data
- calculate an estimate from the frequency table.

Guided practice

The histogram shows the heights of marigold plants. Calculate an estimate for the median height.

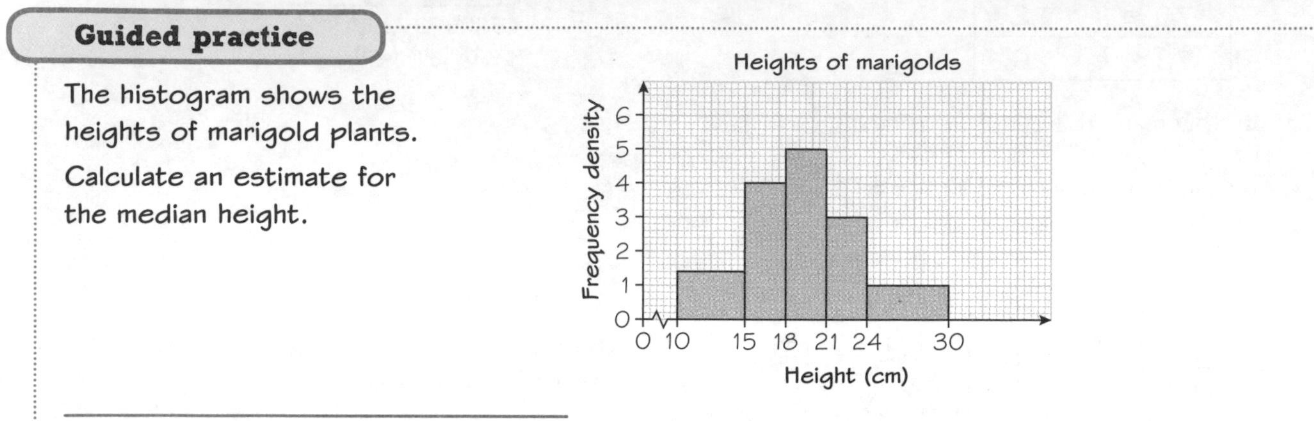

Make a frequency table.

Use the 'start' and 'end' values of each bar, on the horizontal axis.

Height h (cm)	Frequency
$10 \leqslant h < 15$	7
$15 \leqslant h < 18$	12
$18 \leqslant h < 21$	15
$21 \leqslant h < 24$	
$24 \leqslant h < 30$	

19 values

20th to 34th values

Calculate the total frequency:

Total frequency = 49

Median = $\frac{49 + 1}{2}$ = ...th value

Find the class interval containing the 25th value.
$18 \leqslant h < 21$

25th value is $\frac{5}{14}$ = ☐ of the way through the interval.

14 values
5 values
20th value 25th value 34th value
18 19.1 20 21
$\frac{5}{14} \times 3 = 1.1$

Median = 19.1 cm

1 Calculate an estimate for the mean for the histogram in the Guided practice. Give your answer to 1 decimal place. ....

Hint Calculate your estimate from the frequency table.

2 Complete these calculations.

a This class interval contains the 13th to 17th values of a set of data.

The 14th value is .... of the way through the interval.

The interval is .... units wide.

The 14th value is ....

4 values
1 value
13th value 14th value 17th value
26 28
$\frac{1}{4} \times 2 =$ ......

b This class interval contains the 21st to 26th values of a set of data.

.... values

21st value	24th value	26th value

100 — 120

.... × 20 = ...

Find the 24th value. ....

3 The histogram shows the distances people travel to work.

a Complete the frequency table.

Distance, d (km)	Frequency
$0 \leqslant d < 10$	
$10 \leqslant d <$	6
.... $\leqslant d <$	
$20 \leqslant d < 30$	

Distances travelled to work

Frequency density: 0, 1, 2

Distance (km): 0, 5, 10, 15, 20, 25, 30

b Find the modal class. ....

c Estimate the range of the distances. ....

d Calculate the total frequency. ....

e Work out which value is the median. ....

f Find the class interval containing the median. .... $\leqslant d <$ ....

g Find an estimate for the median. .... km

.... values

.... value	.... value	.... value

Exam-style question

4 The histogram shows the ages of horses at a riding stables.

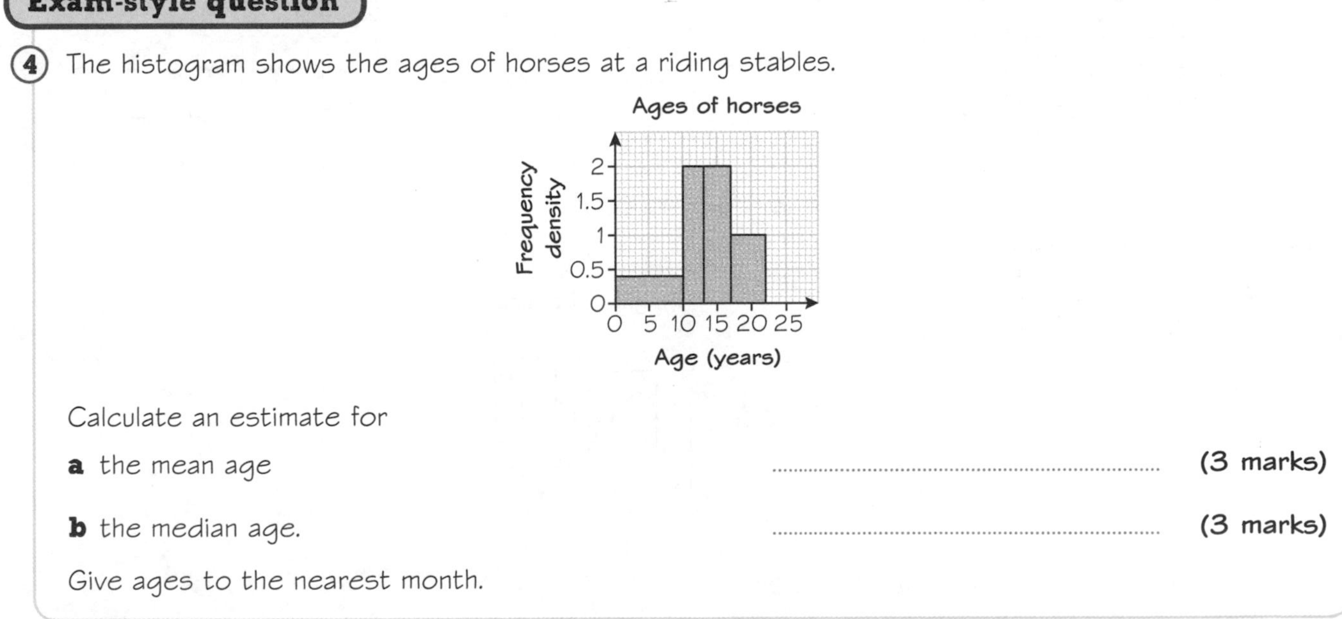

Calculate an estimate for

a the mean age .... **(3 marks)**

b the median age. .... **(3 marks)**

Give ages to the nearest month.

Reflect Why can you only *estimate* the range from a histogram?

Practise the methods

Answer this question to check where to start.

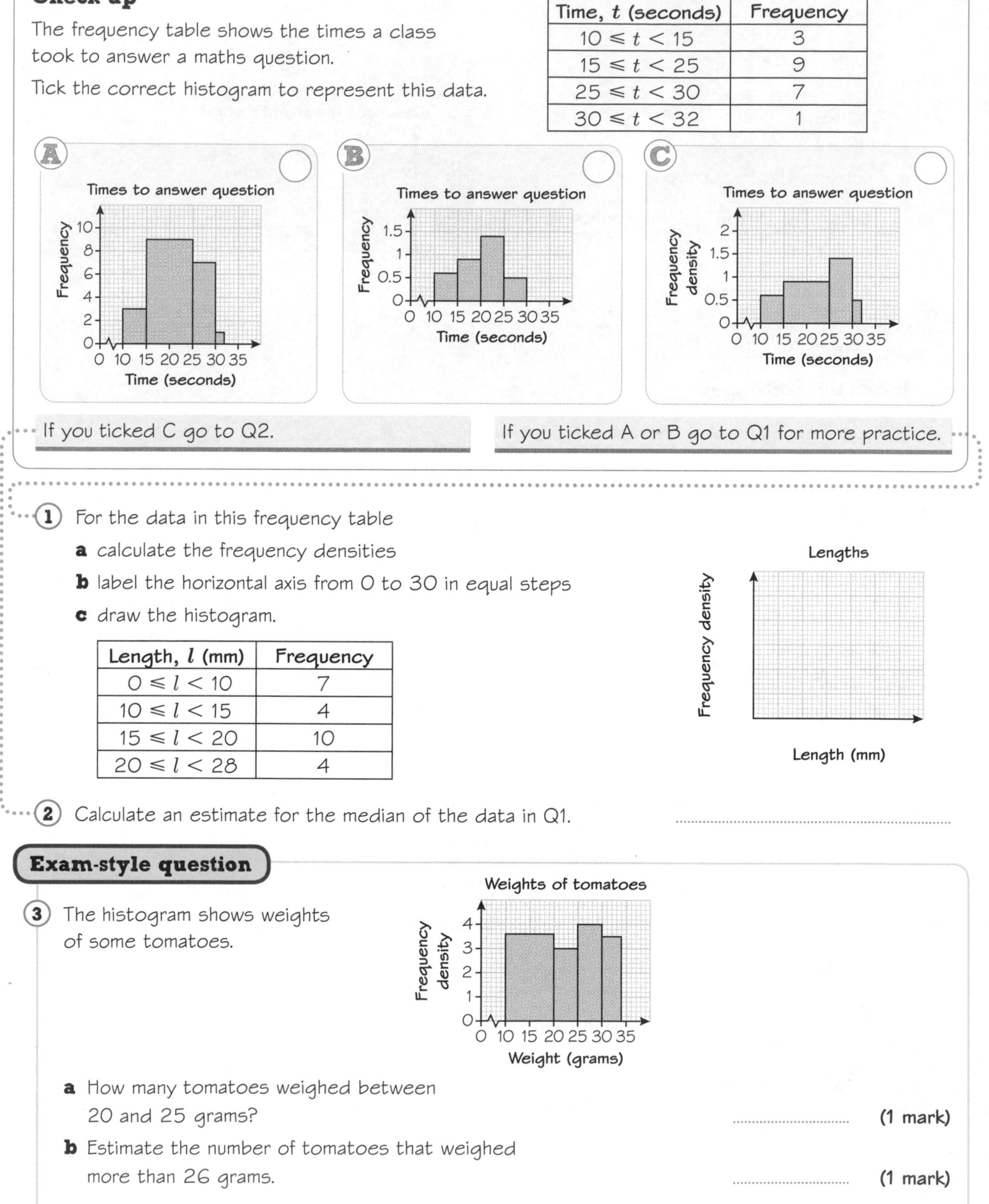

Check up

The frequency table shows the times a class took to answer a maths question.

Tick the correct histogram to represent this data.

Time, t (seconds)	Frequency
$10 \leqslant t < 15$	3
$15 \leqslant t < 25$	9
$25 \leqslant t < 30$	7
$30 \leqslant t < 32$	1

If you ticked C go to Q2. If you ticked A or B go to Q1 for more practice.

1 For the data in this frequency table

a calculate the frequency densities

b label the horizontal axis from 0 to 30 in equal steps

c draw the histogram.

Length, l (mm)	Frequency
$0 \leqslant l < 10$	7
$10 \leqslant l < 15$	4
$15 \leqslant l < 20$	10
$20 \leqslant l < 28$	4

2 Calculate an estimate for the median of the data in Q1. ....

Exam-style question

3 The histogram shows weights of some tomatoes.

a How many tomatoes weighed between 20 and 25 grams? .... **(1 mark)**

b Estimate the number of tomatoes that weighed more than 26 grams. .... **(1 mark)**

c Find the modal class. .... **(2 marks)**

Problem-solve!

Exam-style questions

1. The histogram shows the weights of eggs produced on a farm one day.

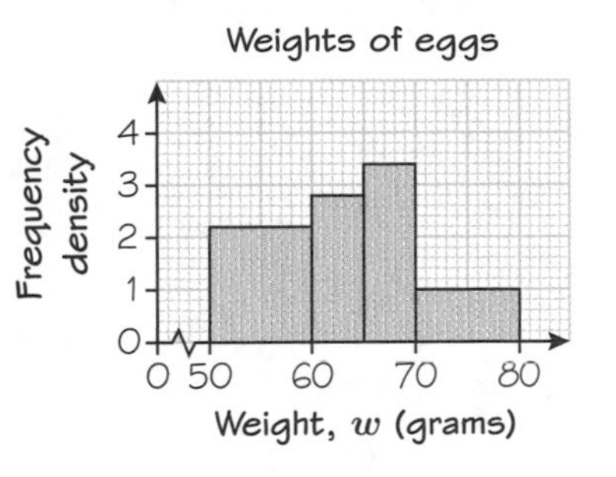

Eggs are sold in four sizes:

Small	Medium	Large	Very large
$w \leqslant 53\,g$	$53\,g < w \leqslant 63\,g$	$63\,g < w \leqslant 73\,g$	$w > 73\,g$

Calculate an estimate for the number of large eggs produced that day. .... **(4 marks)**

2. The incomplete histogram and frequency table show the times to complete a race.

Time, t (minutes)	Frequency
$3 \leqslant t < 5$	2
$5 \leqslant t < 6$	
$6 \leqslant t < 8$	5
$8 \leqslant t < 11$	

a Complete the frequency table. **(2 marks)**

b Complete the histogram.

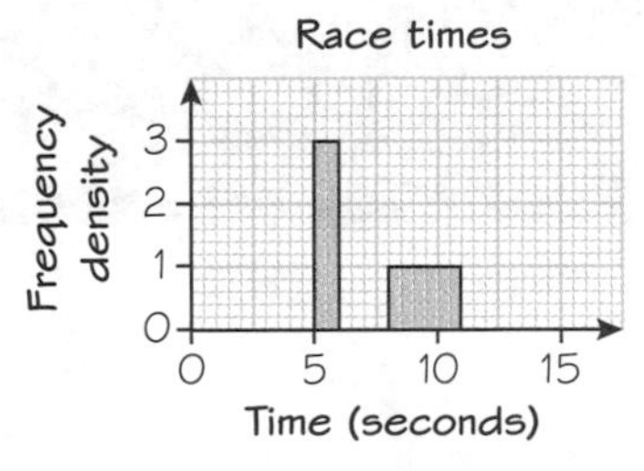

(2 marks)

c Calculate an estimate of the mean time, to the nearest second. .... **(2 marks)**

3. Calculate an estimate for the median of the data in Q2. ....

Now that you have completed this unit, how confident do you feel?

5 Triangles

This unit will help you to find areas, sides, angles and coordinates of triangles.

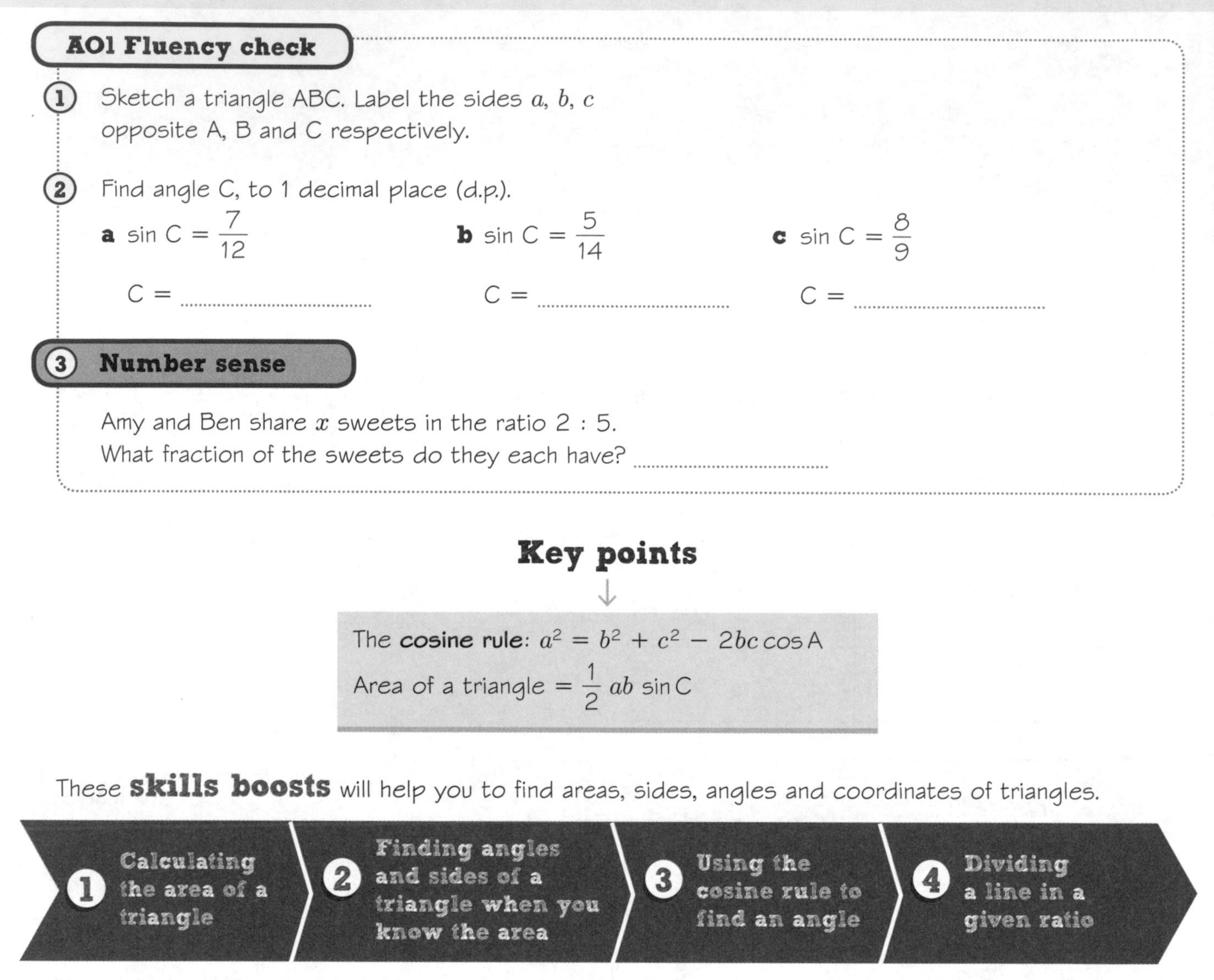

AO1 Fluency check

1. Sketch a triangle ABC. Label the sides a, b, c opposite A, B and C respectively.

2. Find angle C, to 1 decimal place (d.p.).

   a $\sin C = \frac{7}{12}$ C = ...

   b $\sin C = \frac{5}{14}$ C = ...

   c $\sin C = \frac{8}{9}$ C = ...

3. **Number sense**

   Amy and Ben share x sweets in the ratio 2 : 5.
   What fraction of the sweets do they each have? ....

Key points

The **cosine rule**: $a^2 = b^2 + c^2 - 2bc\cos A$

Area of a triangle $= \frac{1}{2}ab\sin C$

These **skills boosts** will help you to find areas, sides, angles and coordinates of triangles.

1. Calculating the area of a triangle
2. Finding angles and sides of a triangle when you know the area
3. Using the cosine rule to find an angle
4. Dividing a line in a given ratio

You might have already done some work on triangles. Before starting the first skills boost, rate your confidence using each concept.

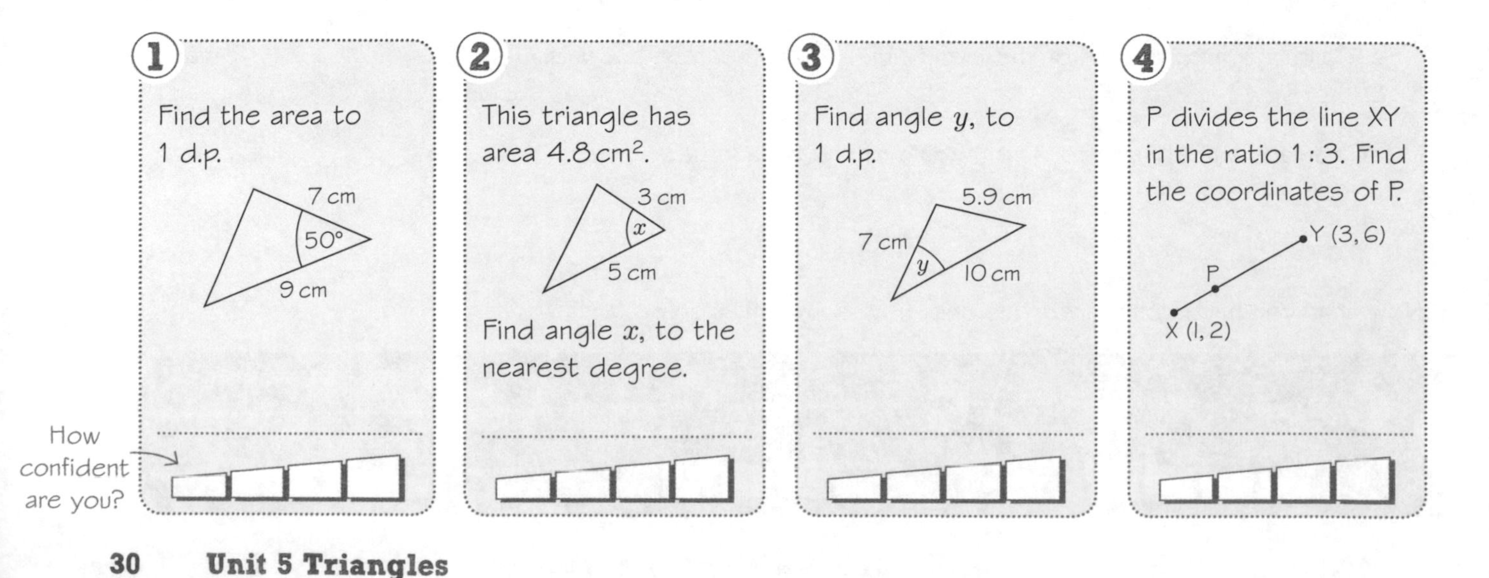

1 Calculating the area of a triangle

When you know two sides and the included angle you can use area $= \frac{1}{2}ab\sin C$ to find the area.

You can rewrite the formula with the letters from your triangle. Replace a and b with the letters for the two sides, and C with the letter for the angle between them.

Area $= \frac{1}{2}ab\sin C$ Area $= \frac{1}{2}yz\sin X$

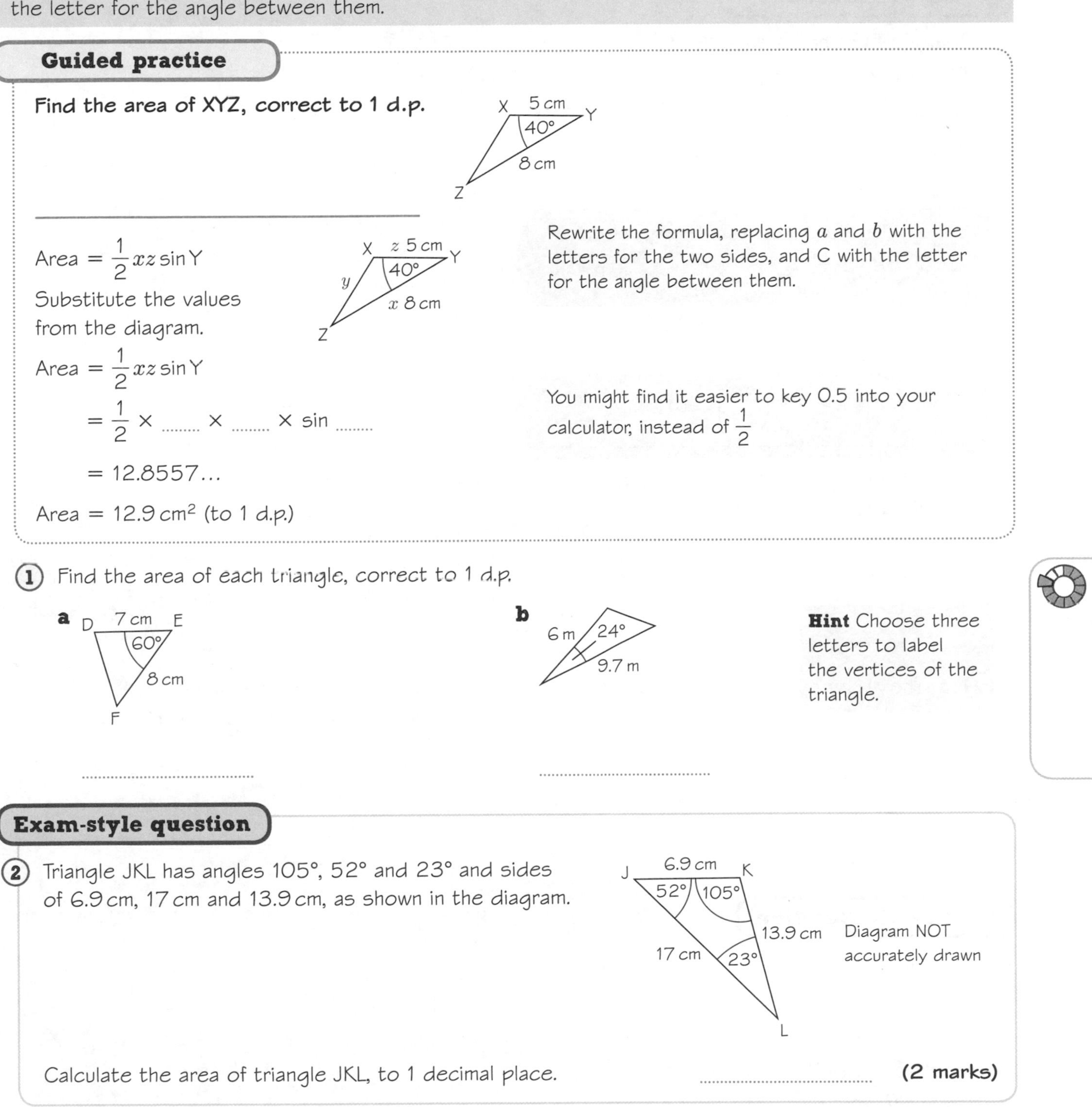

Guided practice

Find the area of XYZ, correct to 1 d.p.

Area $= \frac{1}{2}xz\sin Y$

Rewrite the formula, replacing a and b with the letters for the two sides, and C with the letter for the angle between them.

Substitute the values from the diagram.

Area $= \frac{1}{2}xz\sin Y$

$= \frac{1}{2} \times$ .... $\times$ .... $\times \sin$ ....

You might find it easier to key 0.5 into your calculator, instead of $\frac{1}{2}$

$= 12.8557\ldots$

Area $= 12.9\text{ cm}^2$ (to 1 d.p.)

1. Find the area of each triangle, correct to 1 d.p.

a

....

b

....

Hint Choose three letters to label the vertices of the triangle.

Exam-style question

2. Triangle JKL has angles 105°, 52° and 23° and sides of 6.9 cm, 17 cm and 13.9 cm, as shown in the diagram.

Calculate the area of triangle JKL, to 1 decimal place. .... **(2 marks)**

Reflect For Q2, write two other calculations you could have used. Check that they give the same answer.

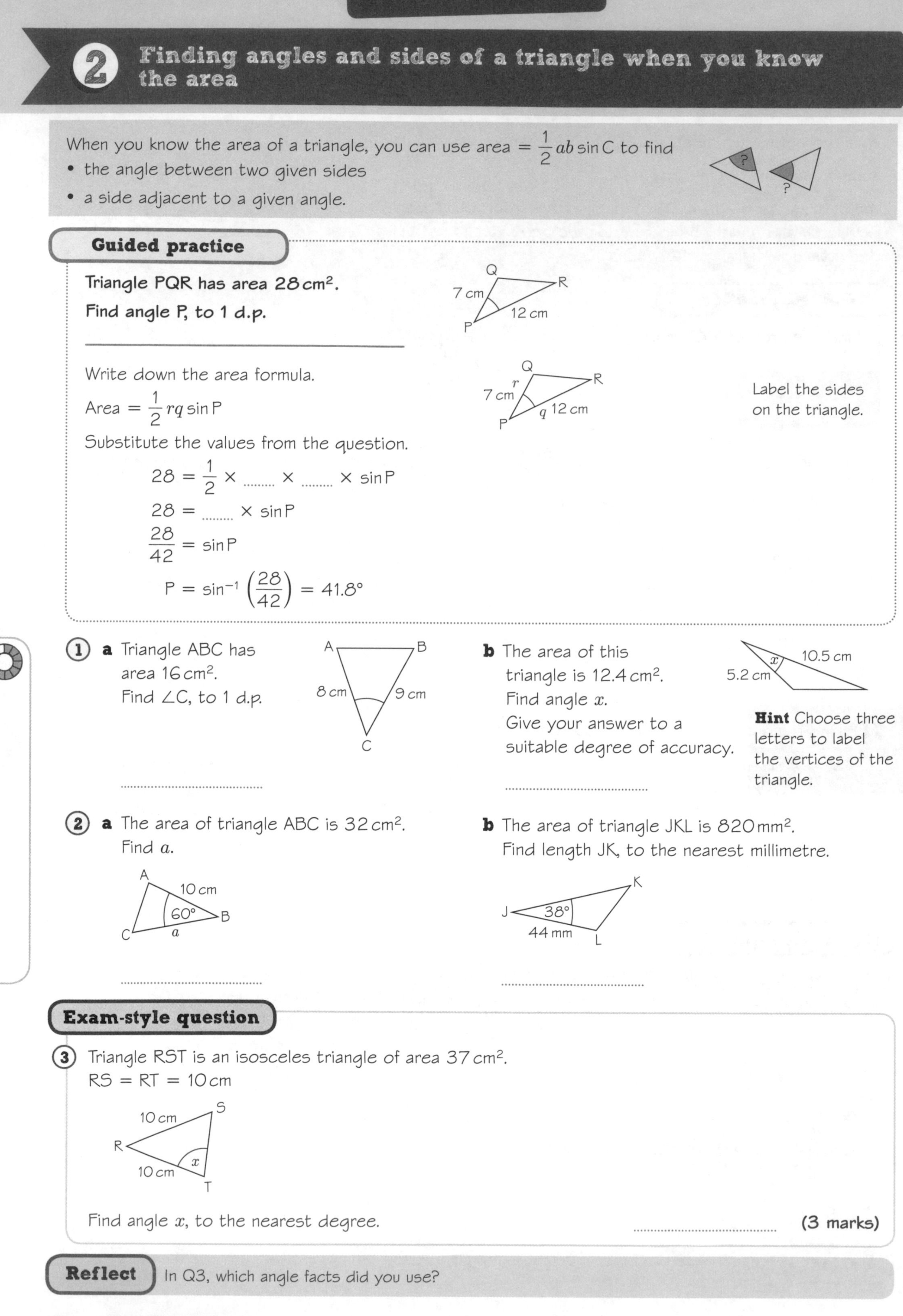

2 Finding angles and sides of a triangle when you know the area

When you know the area of a triangle, you can use area $= \frac{1}{2}ab\sin C$ to find

- the angle between two given sides
- a side adjacent to a given angle.

Guided practice

Triangle PQR has area 28 cm². Find angle P, to 1 d.p.

Write down the area formula.

Area $= \frac{1}{2}rq\sin P$

Label the sides on the triangle.

Substitute the values from the question.

$$28 = \frac{1}{2} \times \text{.........} \times \text{.........} \times \sin P$$

$$28 = \text{.........} \times \sin P$$

$$\frac{28}{42} = \sin P$$

$$P = \sin^{-1}\left(\frac{28}{42}\right) = 41.8°$$

1 **a** Triangle ABC has area 16 cm². Find ∠C, to 1 d.p.

....

b The area of this triangle is 12.4 cm². Find angle x. Give your answer to a suitable degree of accuracy.

Hint Choose three letters to label the vertices of the triangle.

....

2 **a** The area of triangle ABC is 32 cm². Find a.

....

b The area of triangle JKL is 820 mm². Find length JK, to the nearest millimetre.

....

Exam-style question

3 Triangle RST is an isosceles triangle of area 37 cm².
RS = RT = 10 cm

Find angle x, to the nearest degree. .... **(3 marks)**

Reflect In Q3, which angle facts did you use?

3 Using the cosine rule to find an angle

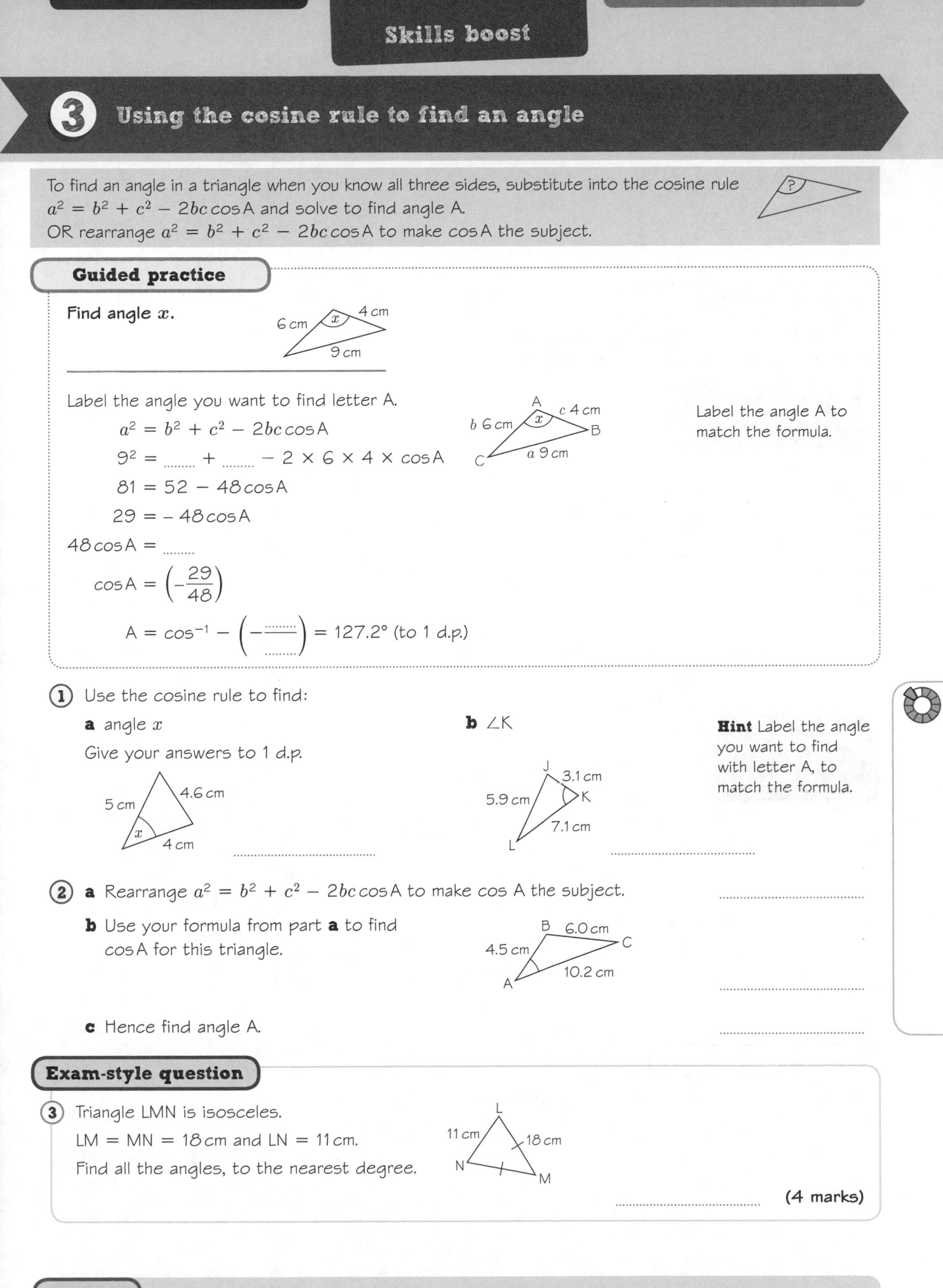

To find an angle in a triangle when you know all three sides, substitute into the cosine rule $a^2 = b^2 + c^2 - 2bc\cos A$ and solve to find angle A.
OR rearrange $a^2 = b^2 + c^2 - 2bc\cos A$ to make $\cos A$ the subject.

Guided practice

Find angle x.

Label the angle you want to find letter A.

$a^2 = b^2 + c^2 - 2bc\cos A$

$9^2 = \dots\dots + \dots\dots - 2 \times 6 \times 4 \times \cos A$

$81 = 52 - 48\cos A$

$29 = -48\cos A$

$48\cos A = \dots\dots$

$\cos A = \left(-\frac{29}{48}\right)$

$A = \cos^{-1} - \left(-\frac{\dots\dots}{\dots\dots}\right) = 127.2°$ (to 1 d.p.)

Label the angle A to match the formula.

1. Use the cosine rule to find:

   a angle x

   Give your answers to 1 d.p.

   ....

   b ∠K

   ....

   Hint Label the angle you want to find with letter A, to match the formula.

2. **a** Rearrange $a^2 = b^2 + c^2 - 2bc\cos A$ to make cos A the subject. ....

   b Use your formula from part **a** to find $\cos A$ for this triangle.

   ....

   c Hence find angle A. ....

Exam-style question

3. Triangle LMN is isosceles.

   LM = MN = 18 cm and LN = 11 cm.

   Find all the angles, to the nearest degree.

   .... **(4 marks)**

Reflect To find angle A, do you prefer to substitute into $a^2 = b^2 + c^2 - 2bc\cos A$ and solve, or to rearrange the formula to make $\cos A$ the subject?

4 Dividing a line in a given ratio

Right-angled triangles on the same sloping line are similar.
ABE and ACD are similar.

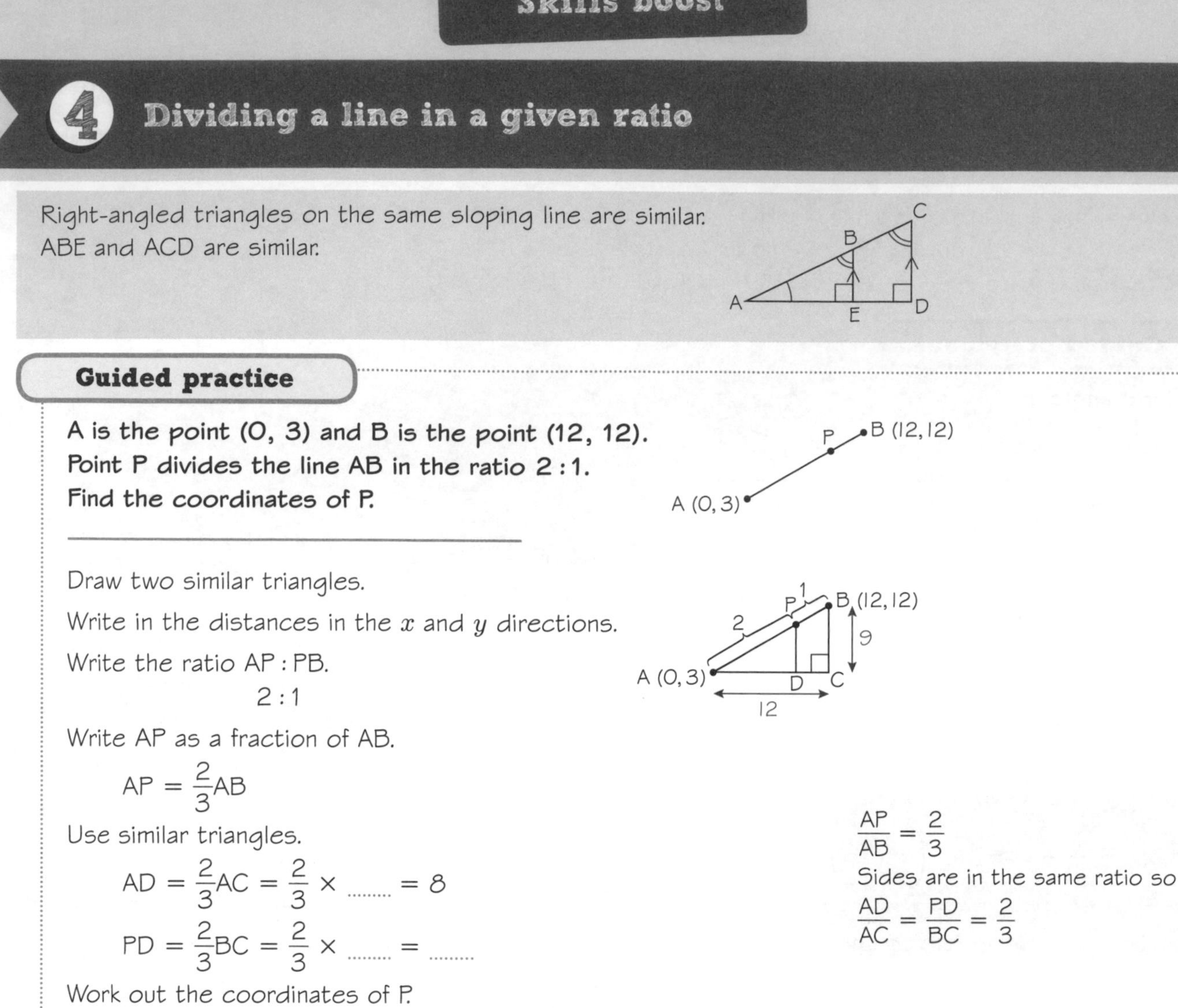

Guided practice

A is the point (0, 3) and B is the point (12, 12).
Point P divides the line AB in the ratio 2 : 1.
Find the coordinates of P.

Draw two similar triangles.

Write in the distances in the x and y directions.

Write the ratio AP : PB.

2 : 1

Write AP as a fraction of AB.

$AP = \frac{2}{3}AB$

$\frac{AP}{AB} = \frac{2}{3}$

Sides are in the same ratio so

$\frac{AD}{AC} = \frac{PD}{BC} = \frac{2}{3}$

Use similar triangles.

$AD = \frac{2}{3}AC = \frac{2}{3} \times$ .... $= 8$

$PD = \frac{2}{3}BC = \frac{2}{3} \times$ .... $=$ ....

Work out the coordinates of P.

P (.....,.....)
6
A (0, 3) 8 D(.....,.....)

P = (8, 9)

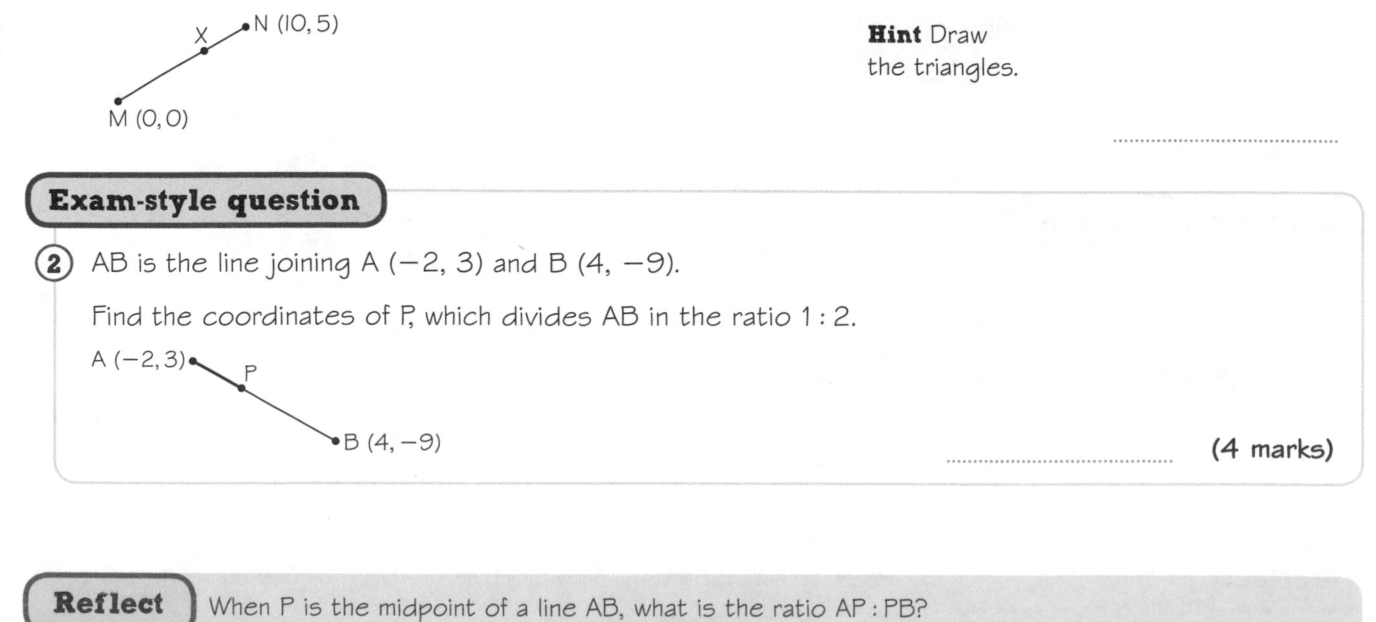

1. M is the point (0, 0) and N is the point (10, 5). Point X divides the line MN in the ratio 3 : 2. Find the coordinates of X.

Hint Draw the triangles.

....

Exam-style question

2. AB is the line joining A (−2, 3) and B (4, −9).

   Find the coordinates of P, which divides AB in the ratio 1 : 2.

.... **(4 marks)**

Reflect When P is the midpoint of a line AB, what is the ratio AP : PB?

Practise the methods

Answer this question to check where to start.

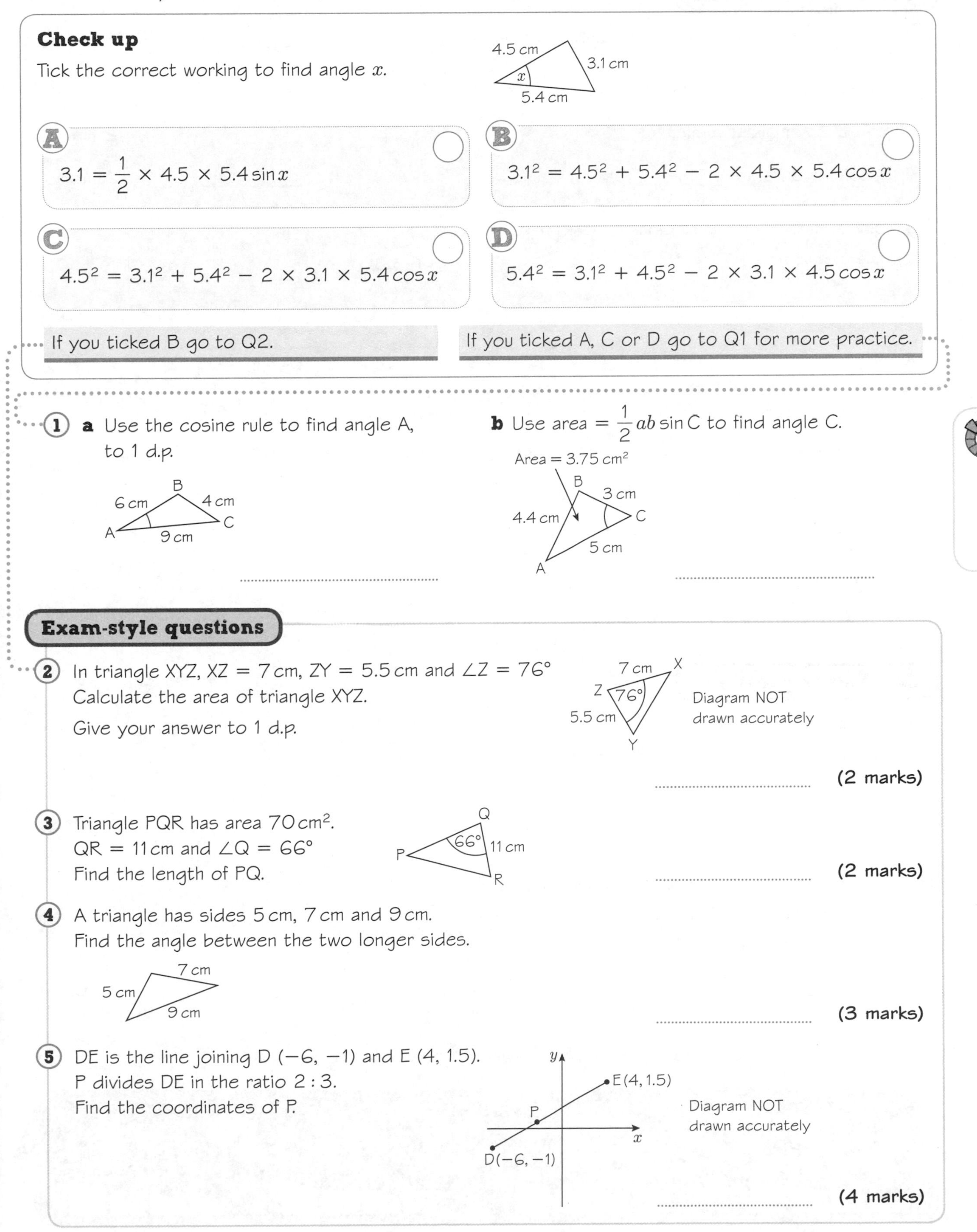

Check up

Tick the correct working to find angle x.

A $3.1 = \frac{1}{2} \times 4.5 \times 5.4 \sin x$

B $3.1^2 = 4.5^2 + 5.4^2 - 2 \times 4.5 \times 5.4 \cos x$

C $4.5^2 = 3.1^2 + 5.4^2 - 2 \times 3.1 \times 5.4 \cos x$

D $5.4^2 = 3.1^2 + 4.5^2 - 2 \times 3.1 \times 4.5 \cos x$

If you ticked B go to Q2.

If you ticked A, C or D go to Q1 for more practice.

1 **a** Use the cosine rule to find angle A, to 1 d.p.

....

b Use area $= \frac{1}{2}ab \sin C$ to find angle C.

....

Exam-style questions

2 In triangle XYZ, XZ = 7 cm, ZY = 5.5 cm and $\angle Z = 76°$

Calculate the area of triangle XYZ.

Give your answer to 1 d.p.

.... **(2 marks)**

3 Triangle PQR has area 70 cm².

QR = 11 cm and $\angle Q = 66°$

Find the length of PQ.

.... **(2 marks)**

4 A triangle has sides 5 cm, 7 cm and 9 cm.

Find the angle between the two longer sides.

.... **(3 marks)**

5 DE is the line joining D (−6, −1) and E (4, 1.5).

P divides DE in the ratio 2 : 3.

Find the coordinates of P.

.... **(4 marks)**

Get back on track

Problem-solve!

Exam-style questions

1. Find the area of an equilateral triangle of side 11 cm. .... **(2 marks)**

2. An isosceles triangle has sides 6.5 cm, 6.5 cm and 4.1 cm.

   a Find its smallest angle. .... **(2 marks)**

   b Calculate its area. .... **(2 marks)**

3. Point R divides the line PQ in the ratio $m:n$.
   Find m and n.

   P(−6, 12)
   R(−3, 6)
   Q(−1, 2)

   .... **(3 marks)**

4. These two triangles both have area 6 cm².

   6 cm
   4 cm
   B

   Triangle 1, ∠B is acute

   4 cm
   B
   6 cm
   A
   C

   Triangle 2, ∠B is obtuse

   a Show that $\sin B = 0.5$ for each triangle. ....

   b Solve the equation $\sin B = 0.5$ to find

   i the acute ∠B in triangle 1 ....

   ii the obtuse ∠B in triangle 2. ....

Exam-style questions

5. In triangle LMN, ∠M is obtuse.
   The area of triangle LMN is 35 cm².
   Find the size of ∠M.

   N
   L
   10 cm
   8 cm
   M

   .... **(3 marks)**

6. A is the point (−3, 6) and P is the point (−2, 4).
   P divides the line AB in the ratio 1 : 3.
   Find the coordinates of B.

   .... **(4 marks)**

Now that you have completed this unit, how confident do you feel?

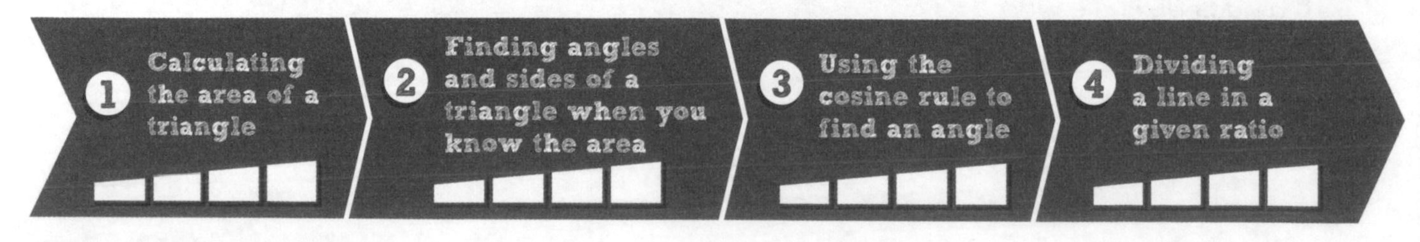

6 Pythagoras and trigonometry in 3D

This unit will help you find missing lengths and angles in 3D solids.

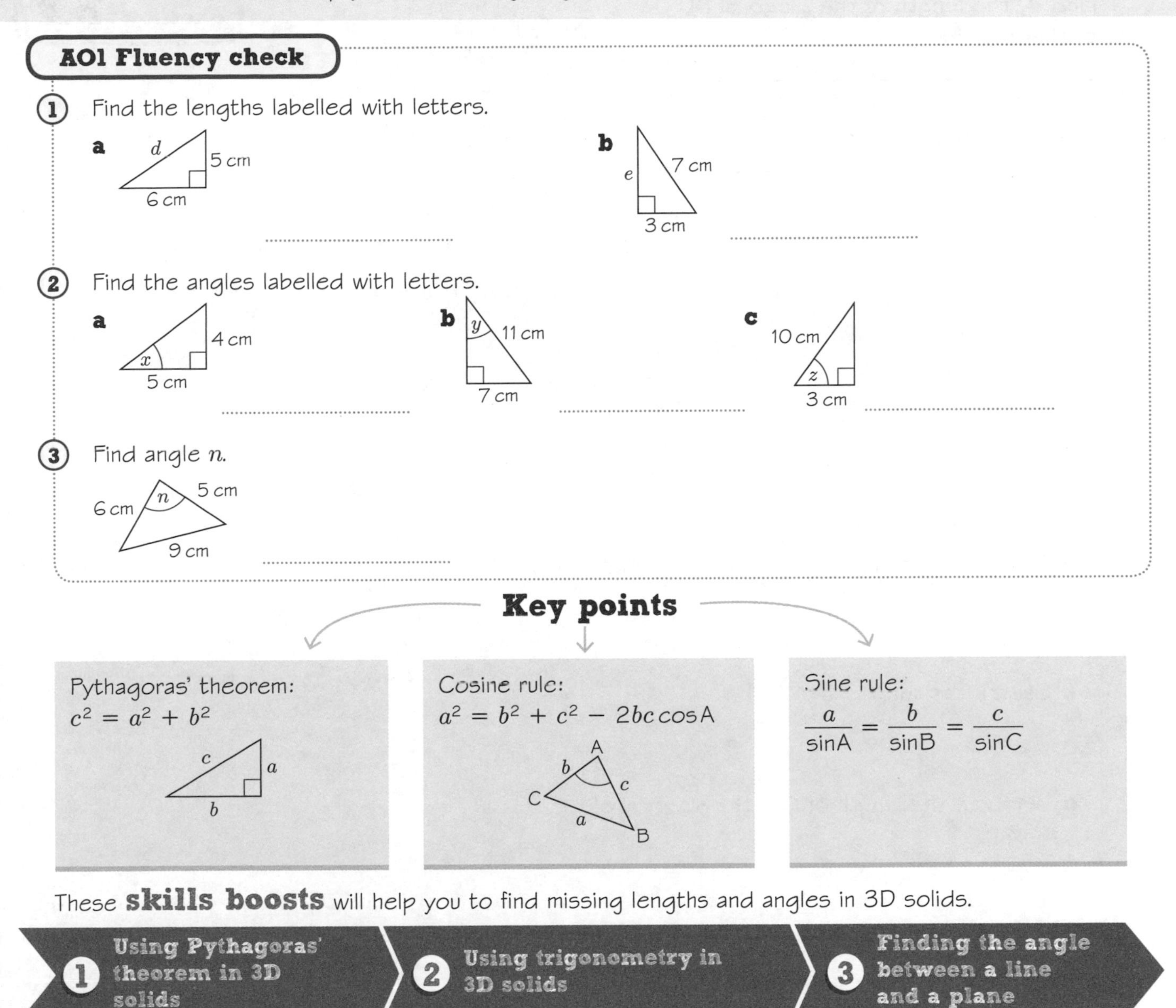

AO1 Fluency check

1 Find the lengths labelled with letters.

a ....

b ....

2 Find the angles labelled with letters.

a ....

b ....

c ....

3 Find angle n.

....

Key points

Pythagoras' theorem:
$c^2 = a^2 + b^2$

Cosine rule:
$a^2 = b^2 + c^2 - 2bc\cos A$

Sine rule:
$\frac{a}{\sin A} = \frac{b}{\sin B} = \frac{c}{\sin C}$

These **skills boosts** will help you to find missing lengths and angles in 3D solids.

1 Using Pythagoras' theorem in 3D solids

2 Using trigonometry in 3D solids

3 Finding the angle between a line and a plane

You might have already done some work on 3D solids. Before starting the first skills boost, rate your confidence using each concept.

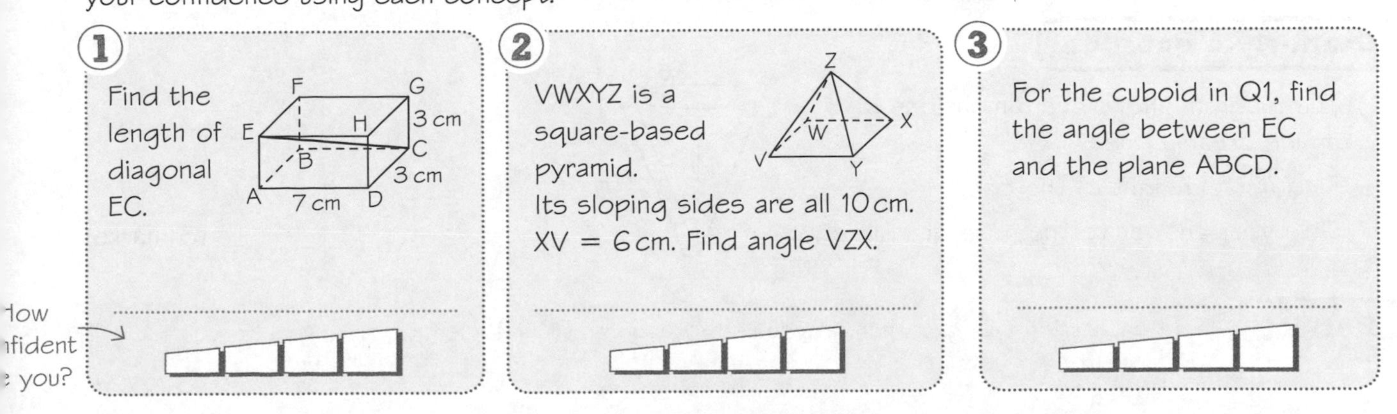

1 Find the length of diagonal EC.

....

2 VWXYZ is a square-based pyramid.
Its sloping sides are all 10 cm.
XV = 6 cm. Find angle VZX.

....

3 For the cuboid in Q1, find the angle between EC and the plane ABCD.

....

How confident are you?

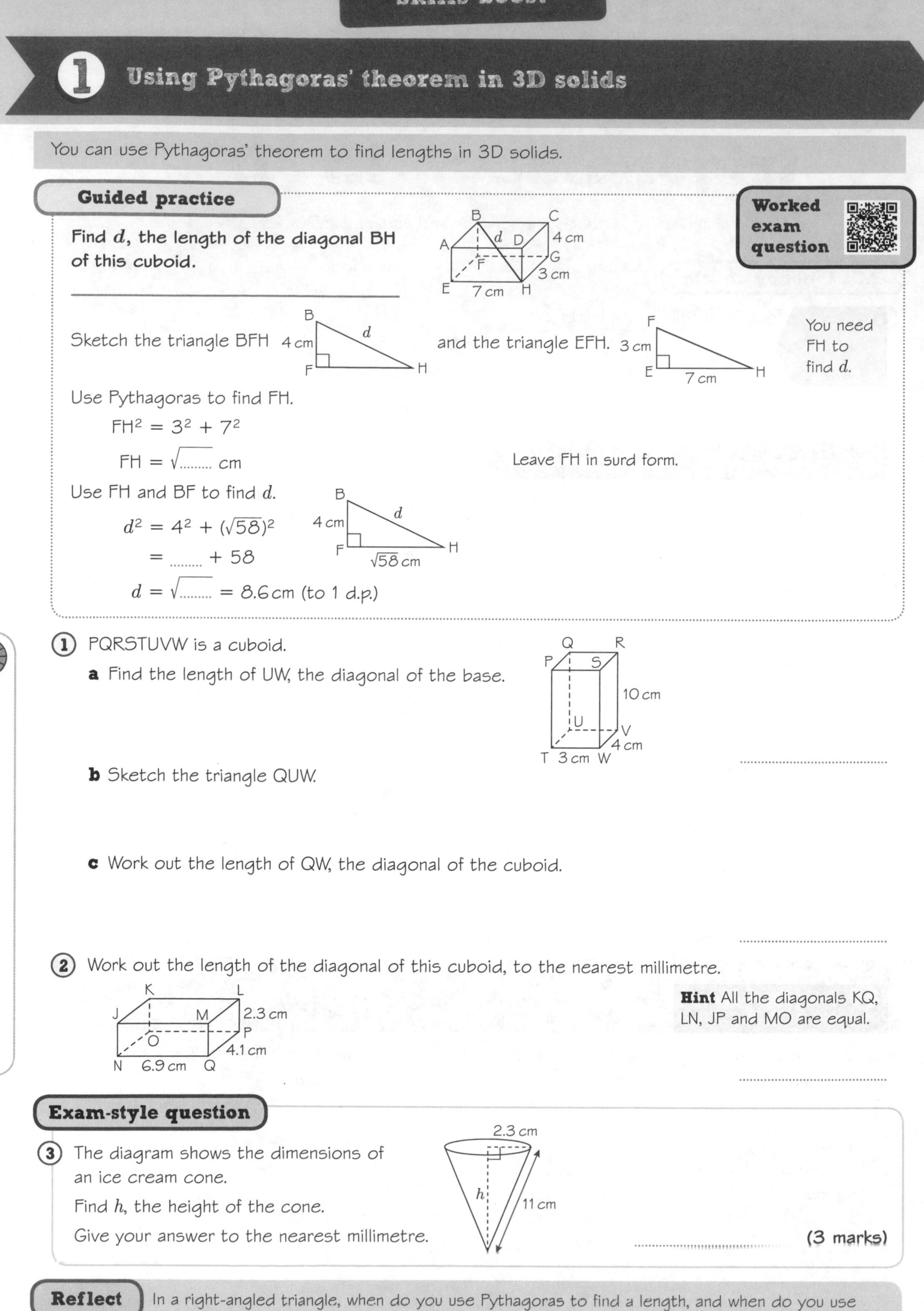

1 Using Pythagoras' theorem in 3D solids

You can use Pythagoras' theorem to find lengths in 3D solids.

Guided practice

Worked exam question

Find d, the length of the diagonal BH of this cuboid.

Sketch the triangle BFH

and the triangle EFH.

You need FH to find d.

Use Pythagoras to find FH.

$FH^2 = 3^2 + 7^2$

$FH = \sqrt{\dots\dots}$ cm

Leave FH in surd form.

Use FH and BF to find d.

$d^2 = 4^2 + (\sqrt{58})^2$

$= \dots\dots + 58$

$d = \sqrt{\dots\dots} = 8.6$ cm (to 1 d.p.)

1. PQRSTUVW is a cuboid.
   a Find the length of UW, the diagonal of the base.

....

   b Sketch the triangle QUW.

   c Work out the length of QW, the diagonal of the cuboid.

....

2. Work out the length of the diagonal of this cuboid, to the nearest millimetre.

Hint All the diagonals KQ, LN, JP and MO are equal.

....

Exam-style question

3. The diagram shows the dimensions of an ice cream cone.
   Find h, the height of the cone.
   Give your answer to the nearest millimetre.

.... **(3 marks)**

Reflect In a right-angled triangle, when do you use Pythagoras to find a length, and when do you use trigonometry?

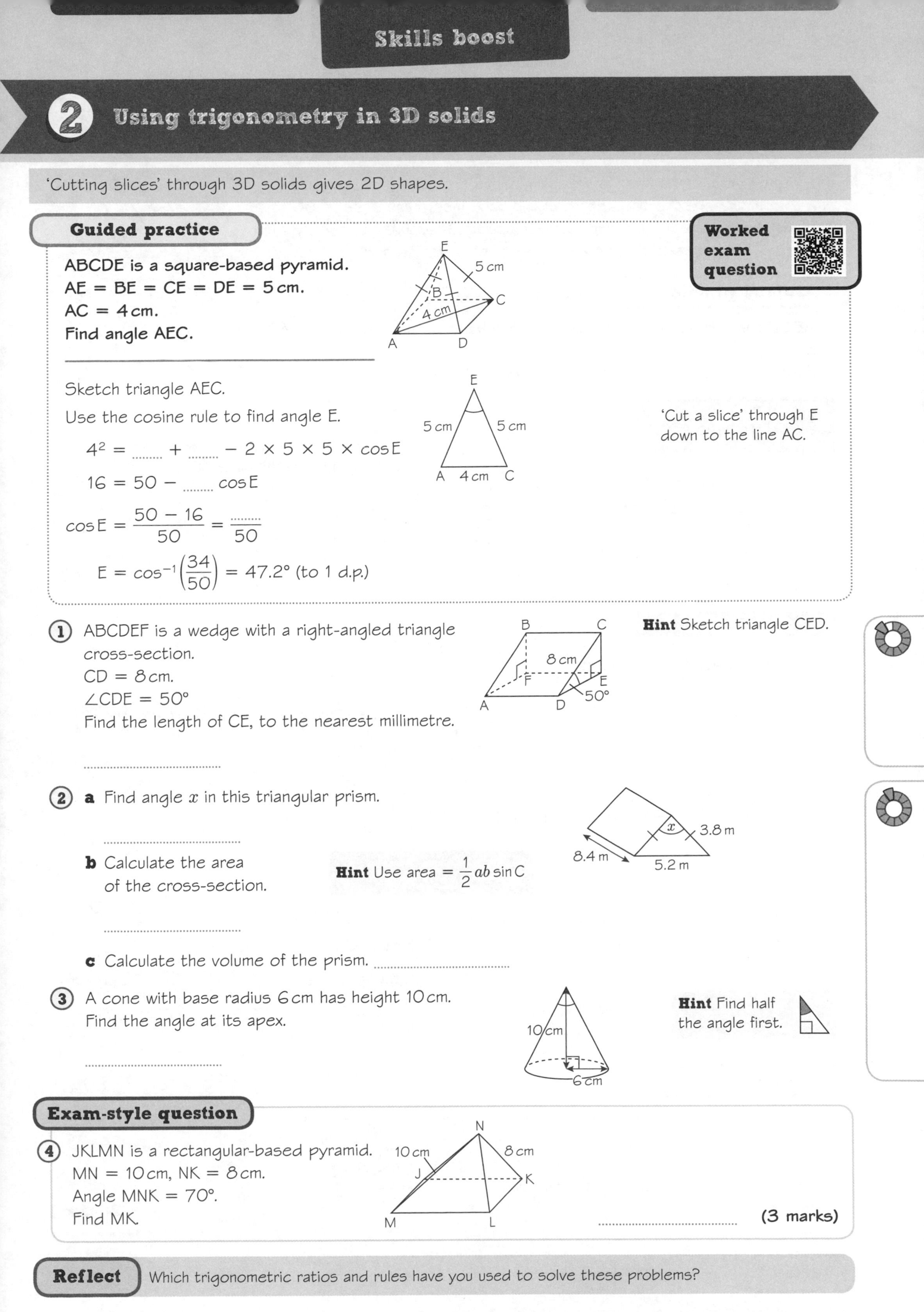

2 Using trigonometry in 3D solids

'Cutting slices' through 3D solids gives 2D shapes.

Guided practice

Worked exam question

ABCDE is a square-based pyramid.
AE = BE = CE = DE = 5 cm.
AC = 4 cm.
Find angle AEC.

Sketch triangle AEC.

Use the cosine rule to find angle E.

'Cut a slice' through E down to the line AC.

$4^2 = \dots + \dots - 2 \times 5 \times 5 \times \cos E$

$16 = 50 - \dots \cos E$

$\cos E = \dfrac{50 - 16}{50} = \dfrac{\dots}{50}$

$E = \cos^{-1}\left(\dfrac{34}{50}\right) = 47.2°$ (to 1 d.p.)

1. ABCDEF is a wedge with a right-angled triangle cross-section.
CD = 8 cm.
∠CDE = 50°
Find the length of CE, to the nearest millimetre.

Hint Sketch triangle CED.

....

2. **a** Find angle x in this triangular prism.

....

b Calculate the area of the cross-section.

Hint Use area $= \frac{1}{2}ab \sin C$

....

c Calculate the volume of the prism. ....

3. A cone with base radius 6 cm has height 10 cm.
Find the angle at its apex.

Hint Find half the angle first.

....

Exam-style question

4. JKLMN is a rectangular-based pyramid.
MN = 10 cm, NK = 8 cm.
Angle MNK = 70°.
Find MK.

.... **(3 marks)**

Reflect Which trigonometric ratios and rules have you used to solve these problems?

3 Finding the angle between a line and a plane

In a **right pyramid** the apex is vertically above the centre of the base.
A line from the apex to the centre of the base meets the base at 90°.
The shaded triangle is in a plane perpendicular to the base of the pyramid.
θ is the angle between the sloping edge of the pyramid and the base.

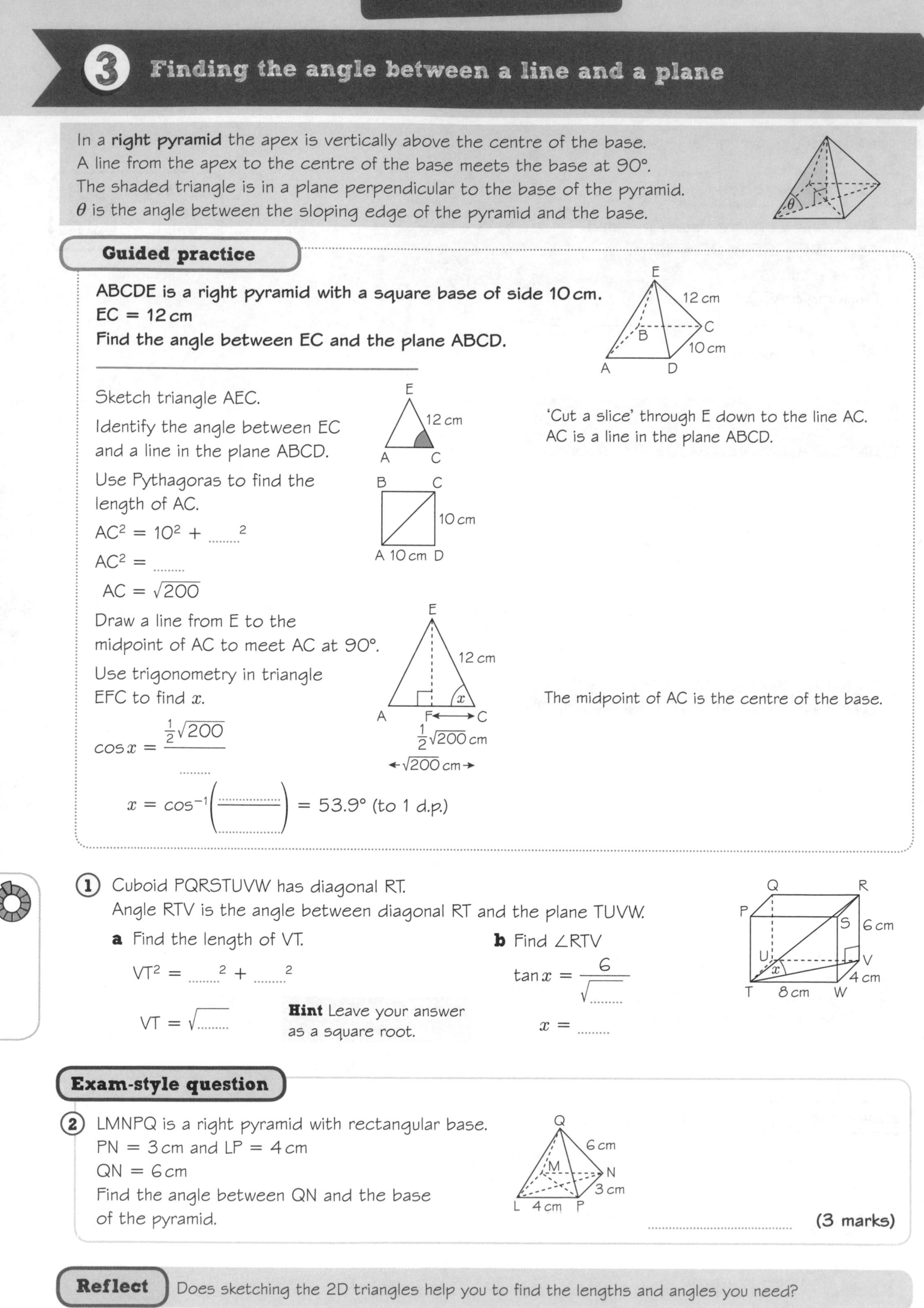

Guided practice

ABCDE is a right pyramid with a square base of side 10 cm.
EC = 12 cm
Find the angle between EC and the plane ABCD.

Sketch triangle AEC.

Identify the angle between EC and a line in the plane ABCD.

'Cut a slice' through E down to the line AC. AC is a line in the plane ABCD.

Use Pythagoras to find the length of AC.

$AC^2 = 10^2 + \text{.......}^2$

$AC^2 = \text{.......}$

$AC = \sqrt{200}$

Draw a line from E to the midpoint of AC to meet AC at 90°.

Use trigonometry in triangle EFC to find x.

The midpoint of AC is the centre of the base.

$\cos x = \dfrac{\frac{1}{2}\sqrt{200}}{\text{.......}}$

$x = \cos^{-1}\left(\dfrac{\text{..........}}{\text{..........}}\right) = 53.9°$ (to 1 d.p.)

1. Cuboid PQRSTUVW has diagonal RT.
   Angle RTV is the angle between diagonal RT and the plane TUVW.
   a Find the length of VT.
   $VT^2 = \text{.......}^2 + \text{.......}^2$
   $VT = \sqrt{\text{.......}}$
   Hint Leave your answer as a square root.
   b Find ∠RTV
   $\tan x = \dfrac{6}{\sqrt{\text{.......}}}$
   $x = \text{.......}$

Exam-style question

2. LMNPQ is a right pyramid with rectangular base.
   PN = 3 cm and LP = 4 cm
   QN = 6 cm
   Find the angle between QN and the base of the pyramid.

   .... **(3 marks)**

Reflect Does sketching the 2D triangles help you to find the lengths and angles you need?

Practise the methods

Answer this question to check where to start.

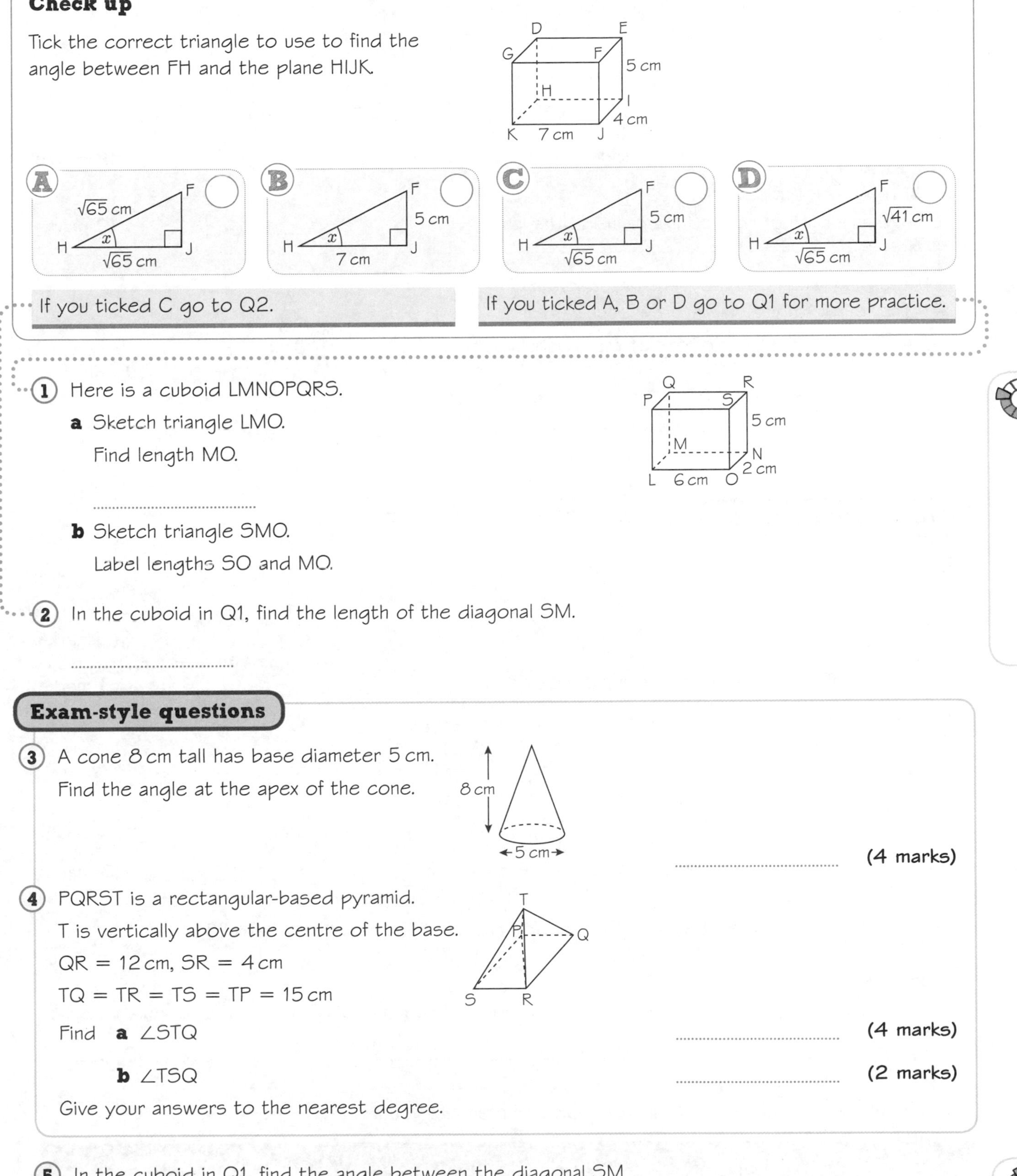

Check up

Tick the correct triangle to use to find the angle between FH and the plane HIJK.

If you ticked C go to Q2.

If you ticked A, B or D go to Q1 for more practice.

1. Here is a cuboid LMNOPQRS.

   a Sketch triangle LMO.

   Find length MO.

   ....

   b Sketch triangle SMO.

   Label lengths SO and MO.

2. In the cuboid in Q1, find the length of the diagonal SM.

   ....

Exam-style questions

3. A cone 8 cm tall has base diameter 5 cm.

   Find the angle at the apex of the cone.

   .... **(4 marks)**

4. PQRST is a rectangular-based pyramid.

   T is vertically above the centre of the base.

   QR = 12 cm, SR = 4 cm

   TQ = TR = TS = TP = 15 cm

   Find **a** ∠STQ .... **(4 marks)**

   b ∠TSQ .... **(2 marks)**

   Give your answers to the nearest degree.

5. In the cuboid in Q1, find the angle between the diagonal SM and the plane LMNO.

   ....

Problem-solve!

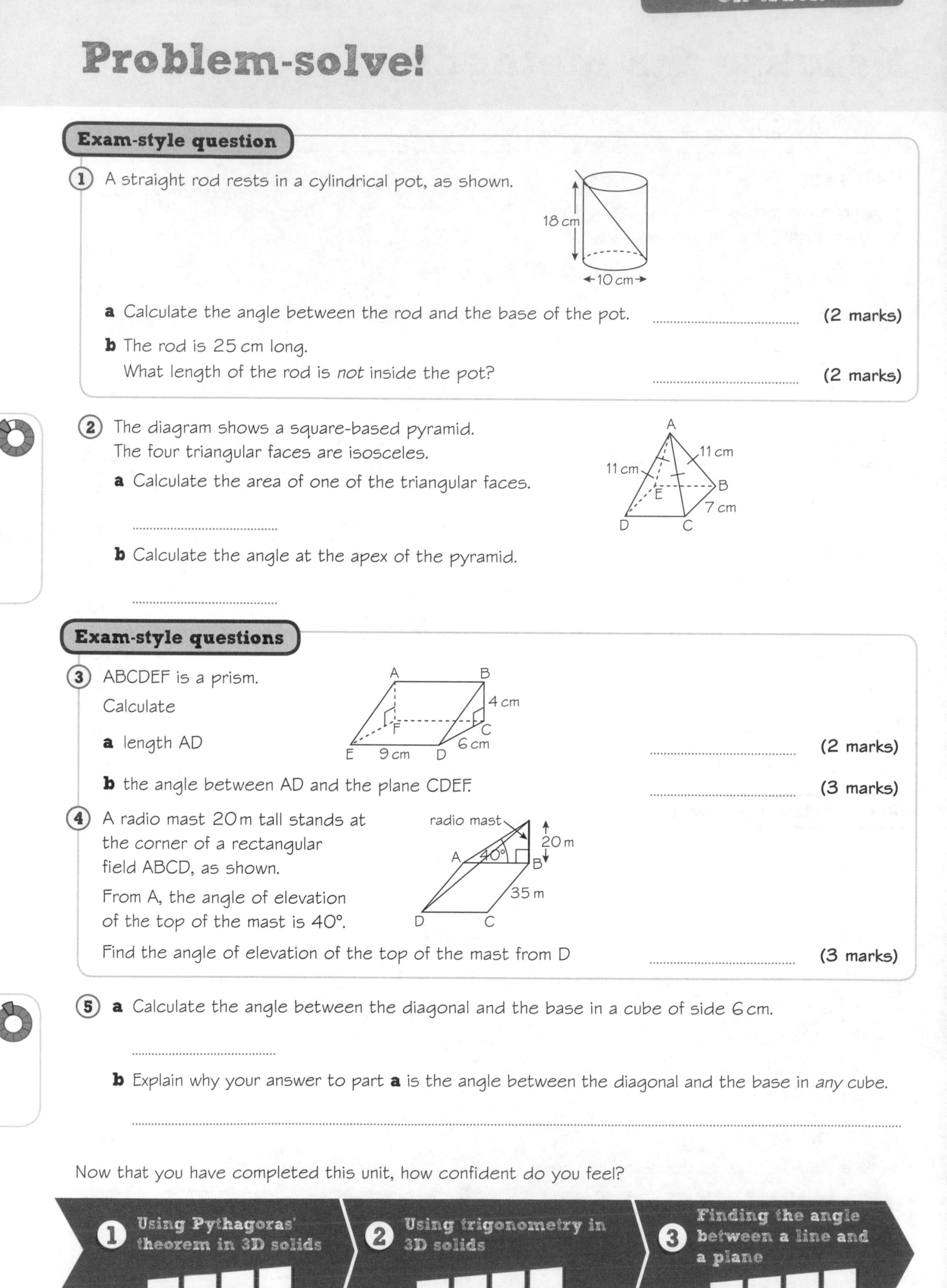

Exam-style question

1. A straight rod rests in a cylindrical pot, as shown.

   a Calculate the angle between the rod and the base of the pot. .... **(2 marks)**

   b The rod is 25 cm long.
   What length of the rod is *not* inside the pot? .... **(2 marks)**

2. The diagram shows a square-based pyramid.
   The four triangular faces are isosceles.

   a Calculate the area of one of the triangular faces.

   ....

   b Calculate the angle at the apex of the pyramid.

   ....

Exam-style questions

3. ABCDEF is a prism.
   Calculate

   a length AD .... **(2 marks)**

   b the angle between AD and the plane CDEF. .... **(3 marks)**

4. A radio mast 20 m tall stands at the corner of a rectangular field ABCD, as shown.
   From A, the angle of elevation of the top of the mast is 40°.
   Find the angle of elevation of the top of the mast from D .... **(3 marks)**

5. **a** Calculate the angle between the diagonal and the base in a cube of side 6 cm.

   ....

   b Explain why your answer to part **a** is the angle between the diagonal and the base in *any* cube.

   ....

Now that you have completed this unit, how confident do you feel?

1. Using Pythagoras' theorem in 3D solids
2. Using trigonometry in 3D solids
3. Finding the angle between a line and a plane

7 3D geometry

This unit will help you to find surface areas and volumes of cones and frustums, and mathematically similar solid shapes.

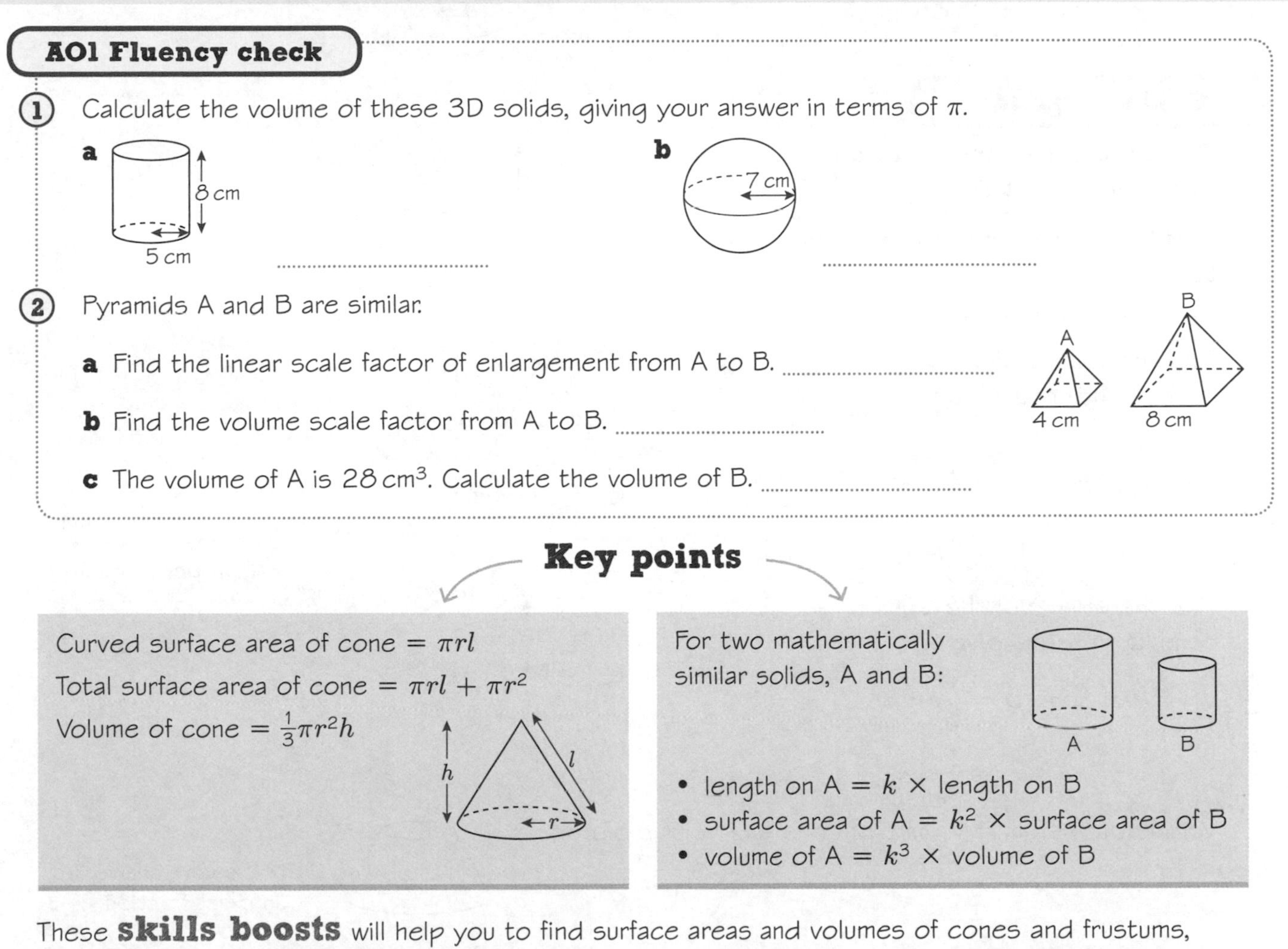

AO1 Fluency check

1. Calculate the volume of these 3D solids, giving your answer in terms of π.

   a (cylinder: 8 cm, 5 cm) ....

   b (sphere: 7 cm) ....

2. Pyramids A and B are similar.

   a Find the linear scale factor of enlargement from A to B. ....

   b Find the volume scale factor from A to B. ....

   c The volume of A is $28\,\text{cm}^3$. Calculate the volume of B. ....

   (A: 4 cm; B: 8 cm)

Key points

Curved surface area of cone $= \pi rl$

Total surface area of cone $= \pi rl + \pi r^2$

Volume of cone $= \frac{1}{3}\pi r^2 h$

For two mathematically similar solids, A and B:

- length on A $= k \times$ length on B
- surface area of A $= k^2 \times$ surface area of B
- volume of A $= k^3 \times$ volume of B

These **skills boosts** will help you to find surface areas and volumes of cones and frustums, and mathematically similar solid shapes.

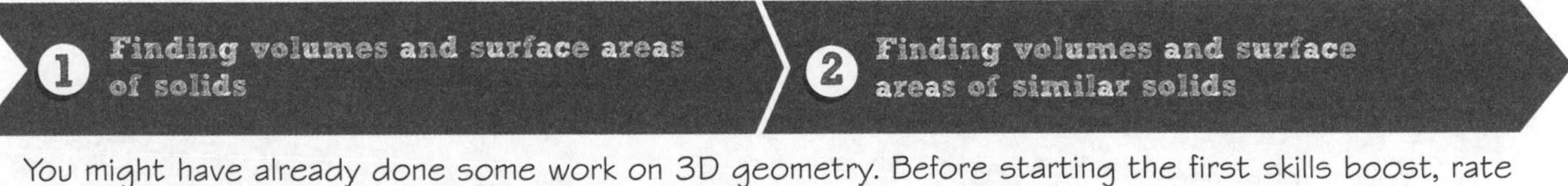

You might have already done some work on 3D geometry. Before starting the first skills boost, rate your confidence using each concept.

1. The top of this cone is cut off to leave a frustum of height 4 cm.

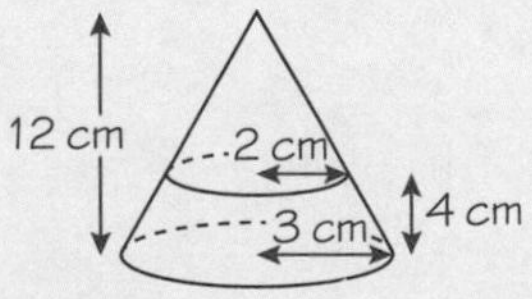

   (12 cm, 2 cm, 3 cm, 4 cm)

   Find the volume of the frustum.

   ....

2. These two cones are mathematically similar.

   (X, Y)

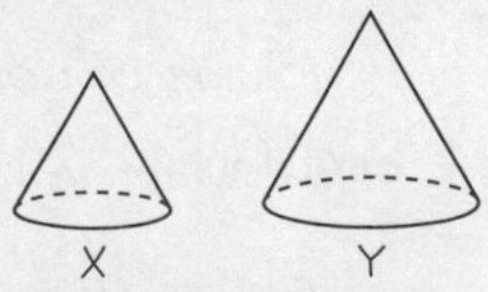

   Cone X has surface area $20\,\text{cm}^2$ and volume $26\,\text{cm}^3$. Cone Y has surface area $30\,\text{cm}^2$. Calculate the volume of cone Y.

   ....

How confident are you?

1 Finding volumes and surface areas of solids

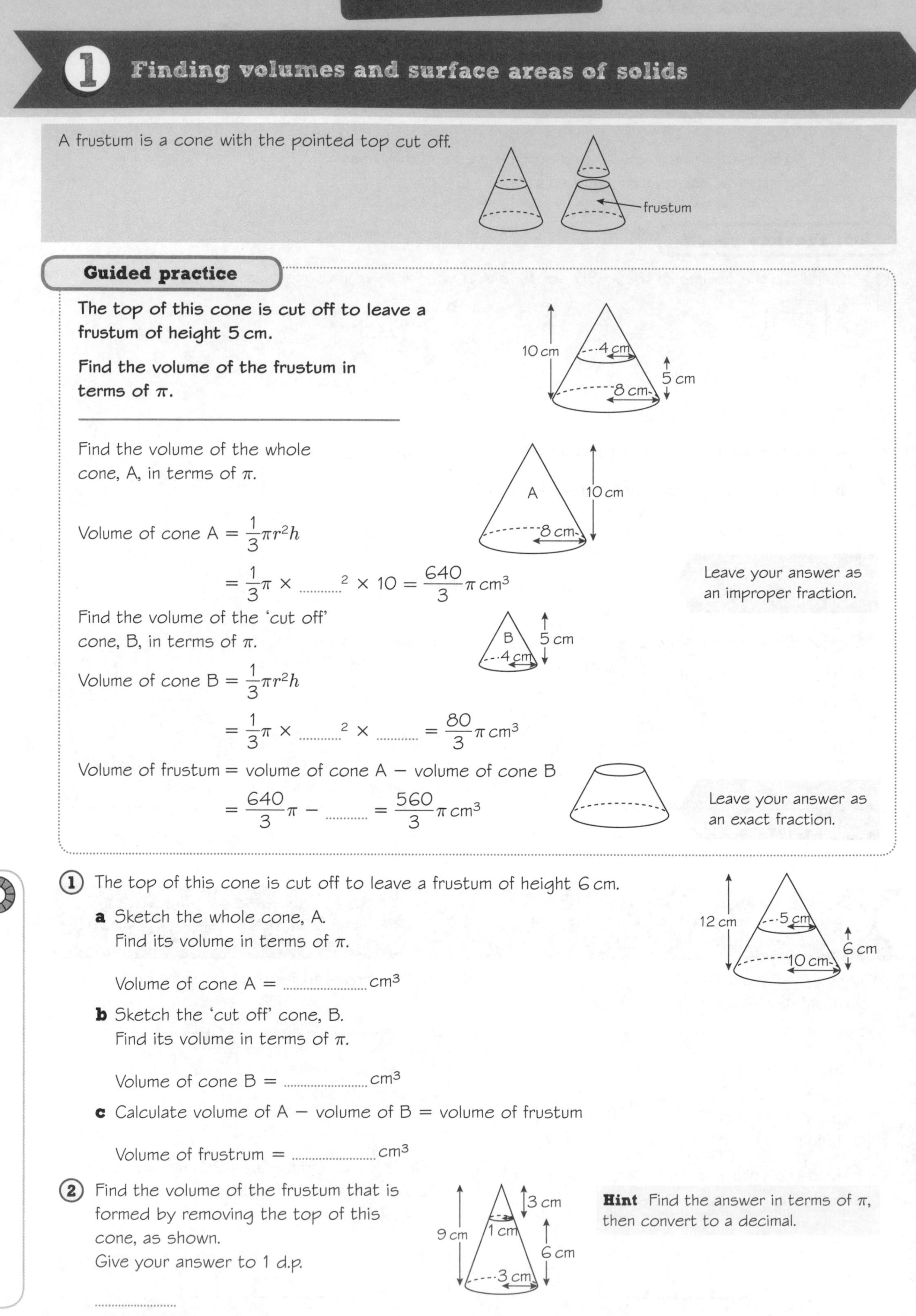

A frustum is a cone with the pointed top cut off.

Guided practice

The top of this cone is cut off to leave a frustum of height 5 cm.

Find the volume of the frustum in terms of π.

Find the volume of the whole cone, A, in terms of π.

Volume of cone A $= \frac{1}{3}\pi r^2 h$

$= \frac{1}{3}\pi \times$ ....$^2 \times 10 = \frac{640}{3}\pi$ cm^3

Leave your answer as an improper fraction.

Find the volume of the 'cut off' cone, B, in terms of π.

Volume of cone B $= \frac{1}{3}\pi r^2 h$

$= \frac{1}{3}\pi \times$ ....$^2 \times$ .... $= \frac{80}{3}\pi$ cm^3

Volume of frustum = volume of cone A − volume of cone B

$= \frac{640}{3}\pi -$ .... $= \frac{560}{3}\pi$ cm^3

Leave your answer as an exact fraction.

1. The top of this cone is cut off to leave a frustum of height 6 cm.

a Sketch the whole cone, A.
Find its volume in terms of π.

Volume of cone A = ... cm^3

b Sketch the 'cut off' cone, B.
Find its volume in terms of π.

Volume of cone B = ... cm^3

c Calculate volume of A − volume of B = volume of frustum

Volume of frustrum = ... cm^3

2. Find the volume of the frustum that is formed by removing the top of this cone, as shown.
Give your answer to 1 d.p.

Hint Find the answer in terms of π, then convert to a decimal.

....

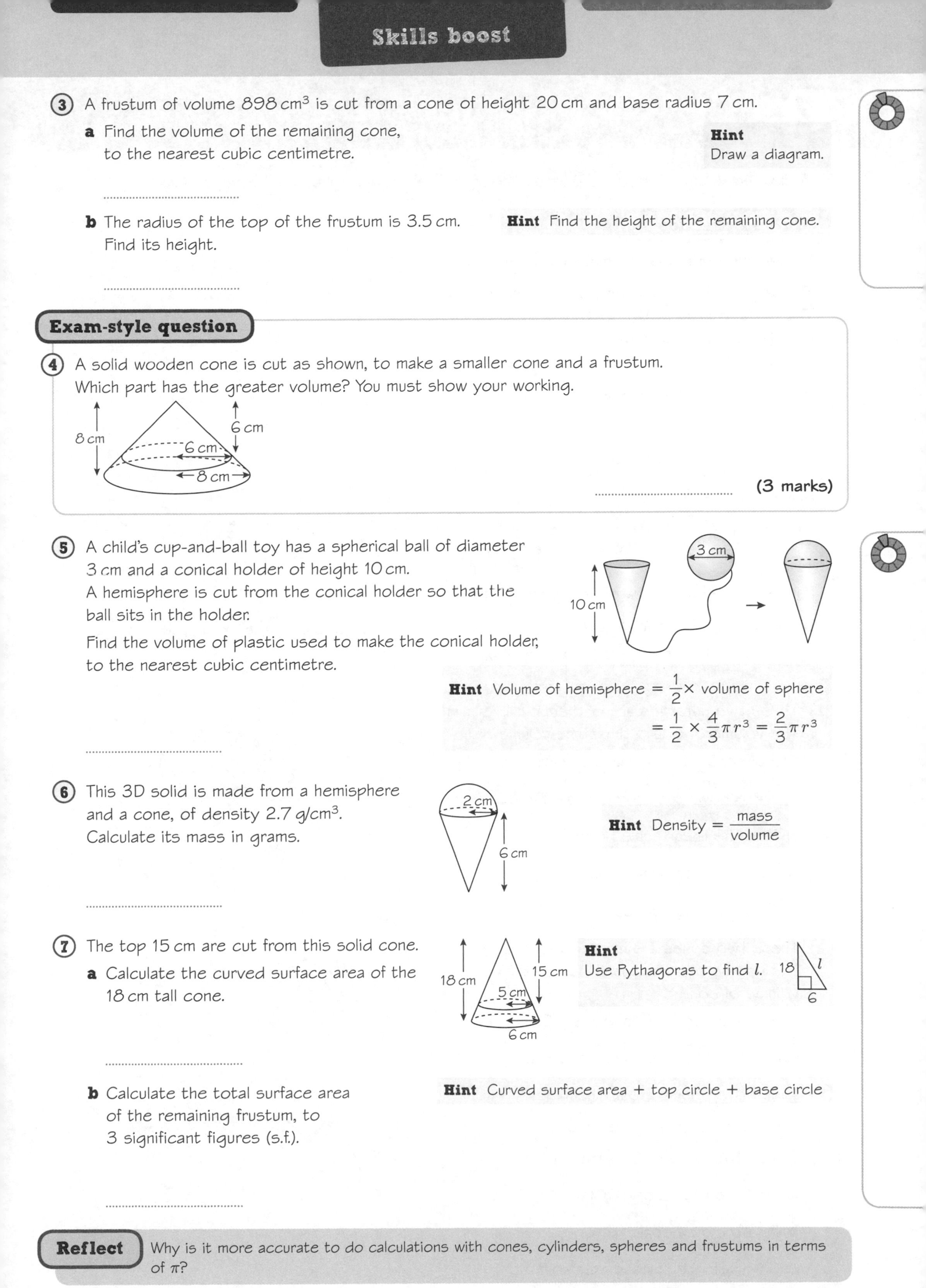

3 A frustum of volume 898 cm³ is cut from a cone of height 20 cm and base radius 7 cm.

a Find the volume of the remaining cone, to the nearest cubic centimetre.

Hint Draw a diagram.

....

b The radius of the top of the frustum is 3.5 cm. Find its height.

Hint Find the height of the remaining cone.

....

Exam-style question

4 A solid wooden cone is cut as shown, to make a smaller cone and a frustum.
Which part has the greater volume? You must show your working.

.... **(3 marks)**

5 A child's cup-and-ball toy has a spherical ball of diameter 3 cm and a conical holder of height 10 cm.
A hemisphere is cut from the conical holder so that the ball sits in the holder.

Find the volume of plastic used to make the conical holder, to the nearest cubic centimetre.

Hint Volume of hemisphere $= \frac{1}{2} \times$ volume of sphere

$$= \frac{1}{2} \times \frac{4}{3}\pi r^3 = \frac{2}{3}\pi r^3$$

....

6 This 3D solid is made from a hemisphere and a cone, of density 2.7 g/cm³.
Calculate its mass in grams.

Hint Density $= \frac{\text{mass}}{\text{volume}}$

....

7 The top 15 cm are cut from this solid cone.

a Calculate the curved surface area of the 18 cm tall cone.

Hint Use Pythagoras to find l.

....

b Calculate the total surface area of the remaining frustum, to 3 significant figures (s.f.).

Hint Curved surface area + top circle + base circle

....

Reflect Why is it more accurate to do calculations with cones, cylinders, spheres and frustums in terms of π?

2 Finding volumes and surface areas of similar solids

To find volumes and surface areas of similar solids, first find the linear scale factor k.

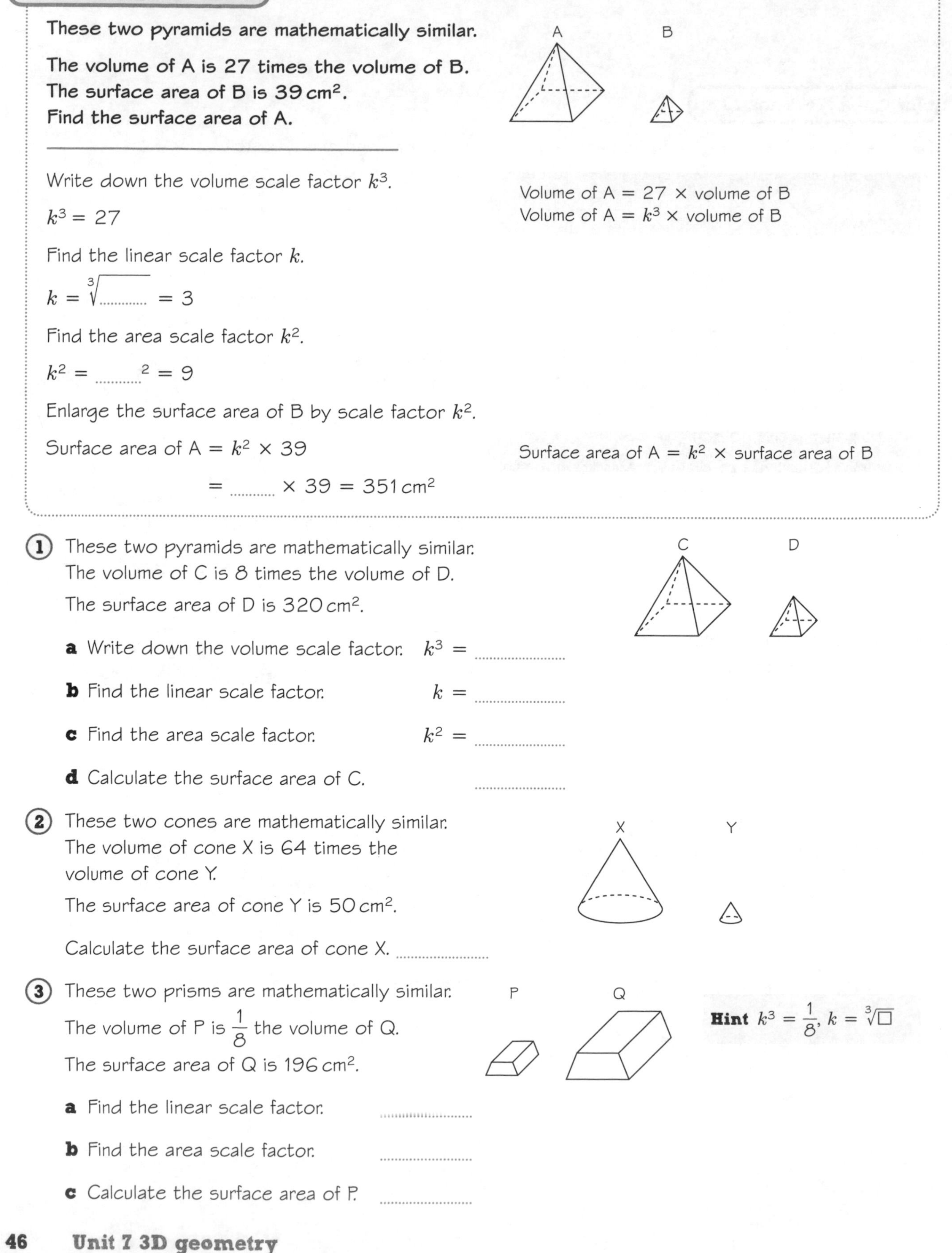

Guided practice

These two pyramids are mathematically similar.

The volume of A is 27 times the volume of B.
The surface area of B is 39 cm².
Find the surface area of A.

Write down the volume scale factor k^3.

$k^3 = 27$

Volume of A = 27 × volume of B
Volume of A = k^3 × volume of B

Find the linear scale factor k.

$k = \sqrt[3]{\dots\dots} = 3$

Find the area scale factor k^2.

$k^2 = \dots\dots^2 = 9$

Enlarge the surface area of B by scale factor k^2.

Surface area of A = k^2 × 39

= .... × 39 = 351 cm²

Surface area of A = k^2 × surface area of B

1. These two pyramids are mathematically similar.
   The volume of C is 8 times the volume of D.
   The surface area of D is 320 cm².

   a Write down the volume scale factor. k^3 = ...

   b Find the linear scale factor. k = ...

   c Find the area scale factor. k^2 = ...

   d Calculate the surface area of C. ....

2. These two cones are mathematically similar.
   The volume of cone X is 64 times the volume of cone Y.
   The surface area of cone Y is 50 cm².

   Calculate the surface area of cone X. ....

3. These two prisms are mathematically similar.
   The volume of P is $\frac{1}{8}$ the volume of Q.
   The surface area of Q is 196 cm².

   Hint $k^3 = \frac{1}{8}$, $k = \sqrt[3]{\square}$

   a Find the linear scale factor. ....

   b Find the area scale factor. ....

   c Calculate the surface area of P. ....

4 Prisms L and M are mathematically similar.
The volume of prism L is $\frac{1}{27}$ the volume of prism M.
The surface area of prism M is $540\,cm^2$.
Calculate the surface area of prism L. ....

Hint Follow the method in Q3.

5 These two pyramids are mathematically similar.
The surface area of pyramid R is 4 times the surface area of pyramid S.
The volume of S is $38\,cm^3$.

R S

a Find the linear scale factor. ....

b Find the volume scale factor. ....

c Calculate the volume of pyramid R. ....

6 Two cylinders are mathematically similar.
Cylinder D has $\frac{1}{9}$ the surface area of cylinder E.
Cylinder E has volume $60\,cm^3$.

Calculate the volume of cylinder D. ....

7 Cone G is mathematically similar to cone H.
The surface area of cone G is $60\,cm^2$.
The volume of cone H is $200\,cm^3$.
The surface area of cone H is $40\,cm^2$.

G H

Hint
Area scale factor = $\frac{\text{area of G}}{\text{area of H}}$

a Find the area scale factor. ....

b Find the volume scale factor. ....

c Calculate the volume of cone G. ....

8 Two prisms are mathematically similar.
Prism I has surface area $80\,mm^2$ and volume $340\,mm^3$.
Prism II has surface area $32\,mm^2$.

Calculate the volume of prism II. ....

Hint Follow the method in Q7.

9 These two hexagonal prisms are mathematically similar.

Prism J has volume $56\,cm^3$.
Prism K has surface area $82\,cm^2$ and volume $98\,cm^3$.

J K

a Find the volume scale factor. ....

b Find the area scale factor. ....

Hint Don't round this value.

c Calculate the surface area of prism J. ....

Exam-style question

10 Two irregularly shaped solids are mathematically similar.
Shape 1 has surface area $152\,cm^2$ and volume $180\,cm^3$.
Shape 2 has volume $270\,cm^3$.
Calculate the surface area of shape 2. .... **(3 marks)**

Reflect How have you used roots and powers in these similarity calculations?

Practise the methods

Answer this question to check where to start.

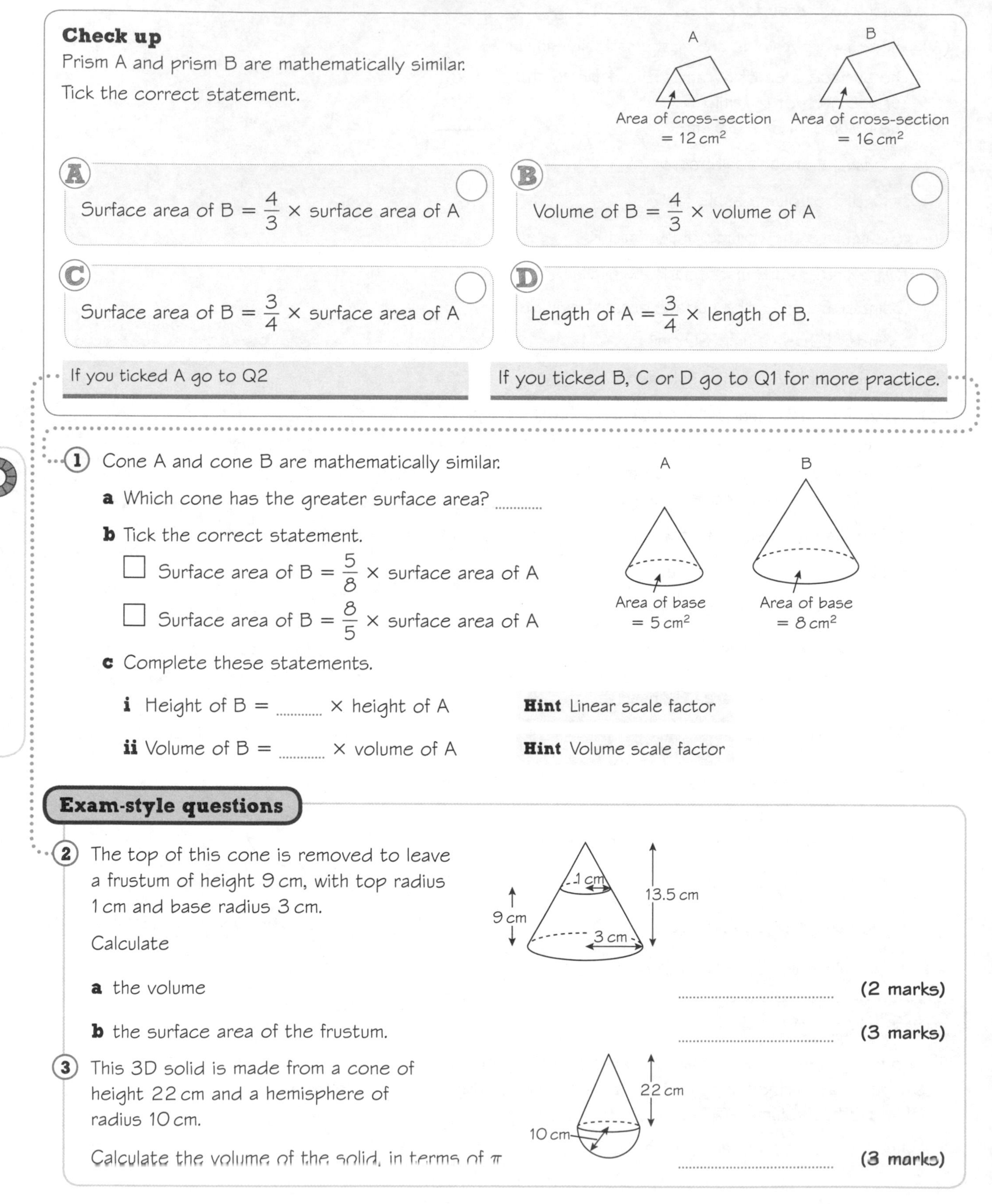

Check up

Prism A and prism B are mathematically similar.

Tick the correct statement.

A Surface area of B = $\frac{4}{3}$ × surface area of A ◯

B Volume of B = $\frac{4}{3}$ × volume of A ◯

C Surface area of B = $\frac{3}{4}$ × surface area of A ◯

D Length of A = $\frac{3}{4}$ × length of B. ◯

If you ticked A go to Q2

If you ticked B, C or D go to Q1 for more practice.

1 Cone A and cone B are mathematically similar.

a Which cone has the greater surface area? ....

b Tick the correct statement.

☐ Surface area of B = $\frac{5}{8}$ × surface area of A

☐ Surface area of B = $\frac{8}{5}$ × surface area of A

c Complete these statements.

i Height of B = ... × height of A **Hint** Linear scale factor

ii Volume of B = ... × volume of A **Hint** Volume scale factor

Exam-style questions

2 The top of this cone is removed to leave a frustum of height 9 cm, with top radius 1 cm and base radius 3 cm.

Calculate

a the volume .... **(2 marks)**

b the surface area of the frustum. .... **(3 marks)**

3 This 3D solid is made from a cone of height 22 cm and a hemisphere of radius 10 cm.

Calculate the volume of the solid, in terms of π .... **(3 marks)**

Problem-solve!

1 The top of a cone has been cut off to leave this frustum.

The sketch shows the original cone.
The red triangle is half the cross-section.

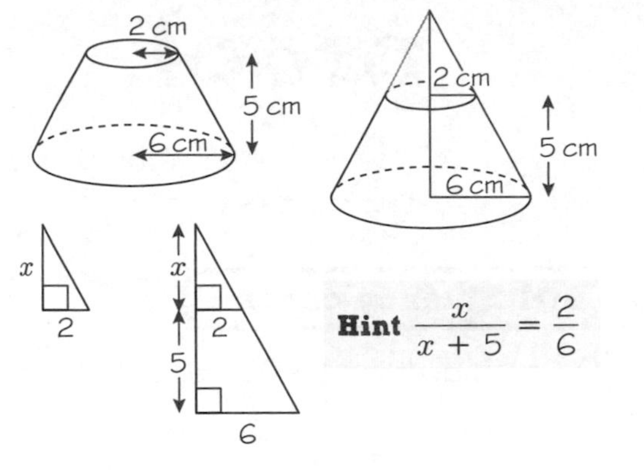

a Use the properties of similar triangles to find the height of the original cone.

Hint $\dfrac{x}{x+5} = \dfrac{2}{6}$

....

b Hence find the volume of the frustum, in terms of π.

....

Exam-style questions

2 A frustum of a cone has top radius 4 cm, base radius 8 cm and height 11 cm.

Calculate its volume.
Give your answer to 3 significant figures.

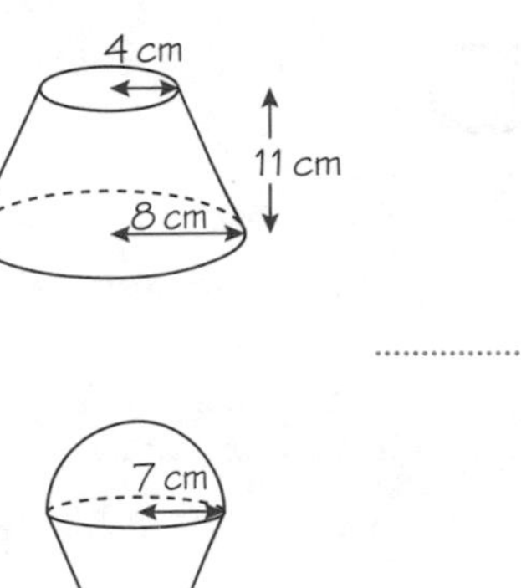

.... **(4 marks)**

3 This solid is formed from a hemisphere of radius 7 cm and a cone.
Its total volume is 2206 cm³.

Calculate the perpendicular height of the cone, to 1 decimal place.

.... **(4 marks)**

4 A water trough is in the shape of a pyramid with the apex cut off to leave a frustum.

By considering the dimensions of the original pyramid, calculate the capacity of the trough, in litres.

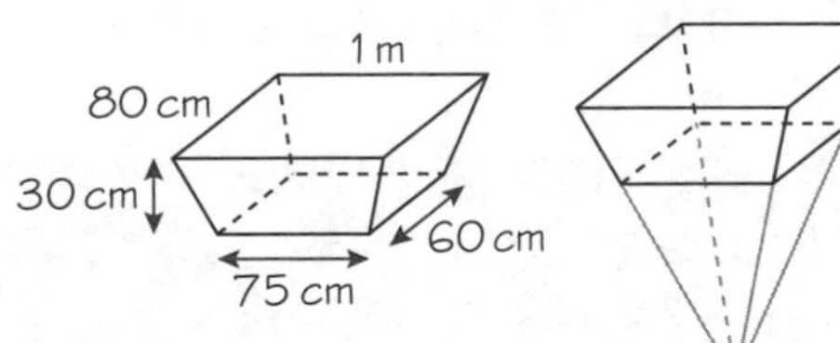

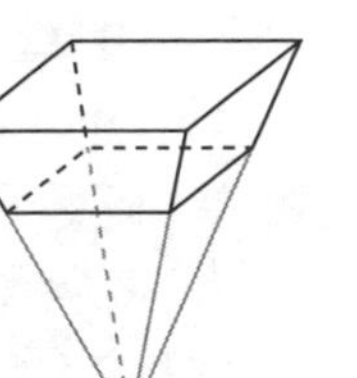

....

5 Calculate the capacity of this wastepaper bin, in cubic centimetres.

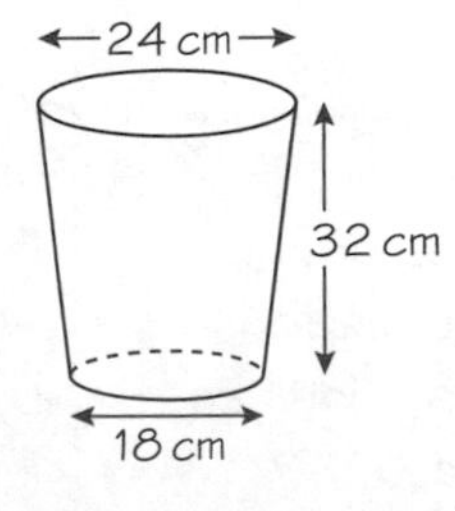

....

Now that you have completed this unit, how confident do you feel?

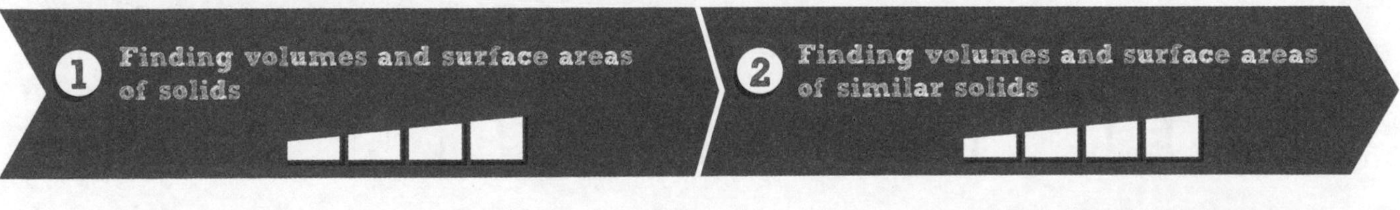

8 Algebraic and geometric proof

This unit will help you to answer questions that ask you to 'prove' or 'show that' a statement is true.

AO1 Fluency check

1 **a** Show that △ABE and △ACD are similar.

A, B, C, D, E; 6 cm; 3 cm; 5 cm; 40°; 60°

b Find the length of CD.

2 Number sense

Here are the general terms of some sequences. Which general terms generate a sequence of

$2n$	$2n - 1$	$2n + 1$	$2(n + 1)$

a odd numbers

b even numbers?

Key points

Two shapes are similar if they have equal angles and their corresponding sides are in the same ratio.

Two triangles are congruent when one of these conditions of congruence is true: SSS SAS AAS RHS

These **skills boosts** will help you to answer questions that ask you to 'prove' or 'show that' a statement is true.

1 Proving results about odd numbers, even numbers and squares
2 Proving geometric results using similarity
3 Proving geometric results using congruence
4 Proving circle theorems

You might have already done some work on proof. Before starting the first skills boost, rate your confidence using each concept.

1 Prove that the product of two odd numbers is always odd.

2 Prove that any two right-angled isosceles triangles are similar.

3 Show that the diagonal AC divides this symmetrical arrowhead into two congruent triangles.

A, C, D, D

4 Prove that an exterior angle to a cyclic quadrilateral is equal to the opposite interior angle.

How confident are you?

1 Proving results about odd numbers, even numbers and squares

You can represent
- consecutive integers as $n - 1, n, n + 1, n + 2, n + 3, \ldots$
- consecutive even numbers as $2n - 2, 2n, 2n + 2, 2n + 4, \ldots$
- consecutive odd numbers as $2n - 1, 2n + 1, 2n + 3, \ldots$

Guided practice

Show that the sum of the squares of two consecutive even numbers is a multiple of 4.

Write the expressions for two consecutive even numbers.

$2n$ and $2n + 2$

$2n$ is even.
$2n + 1$ is odd.
$2n + 2$ is the next even number.

Square each expression.

$(2n)^2$ $\quad$ $(2n + 2)^2$

$= \ldots\ldots$ $\quad$ $= \ldots\ldots + 8n + 4$

Add to find the sum , then factorise.

$(2n)^2 + (2n + 2)^2 = \ldots\ldots + \ldots\ldots + 8n + 4$

$= 8n^2 + 8n + 4$

$= 4(2n^2 + \ldots\ldots + \ldots\ldots)$

This is a multiple of 4.

$4 \times$ any number is a multiple of 4.

1. Show that the difference between the squares of two consecutive even numbers is a multiple of 4.
   Hint Subtract to find the difference, then factorise.
   Hint $(2n)^2$ $\quad$ $(2n + 2)^2$
   ....
2. Prove that the sum of any three consecutive integers is always a multiple of 3.
   Hint Use $n - 1$, n and $n + 1$ for the integers.
   ....
3. Prove that the sum of any two even numbers is always even.
   Hint Let the numbers be $2m$ and $2n$.
   ....
4. Prove that the difference between any two odd numbers is always even.
   ....
5. Show that the sum of the squares of two consecutive odd numbers is always even.
   Hint Use $2n - 1$ and $2n + 1$ for the odd numbers. Show that the sum is a multiple of 2.
   ....

Exam-style question

6. **a** Calculate the difference between 7^2 and 9^2. .... **(1 mark)**

   b Show that the difference between the squares of two consecutive odd numbers is a multiple of 8. .... **(2 marks)**

Reflect Repeat Q2 using n, $n + 1$ and $n + 2$ for the three consecutive integers. Can you still prove the result?

2 Proving geometric results using similarity

For two mathematically similar solids, A and B
- length on A = k × length on B
- surface area of A = k^2 × surface area of B
- volume of A = k^3 × volume of B

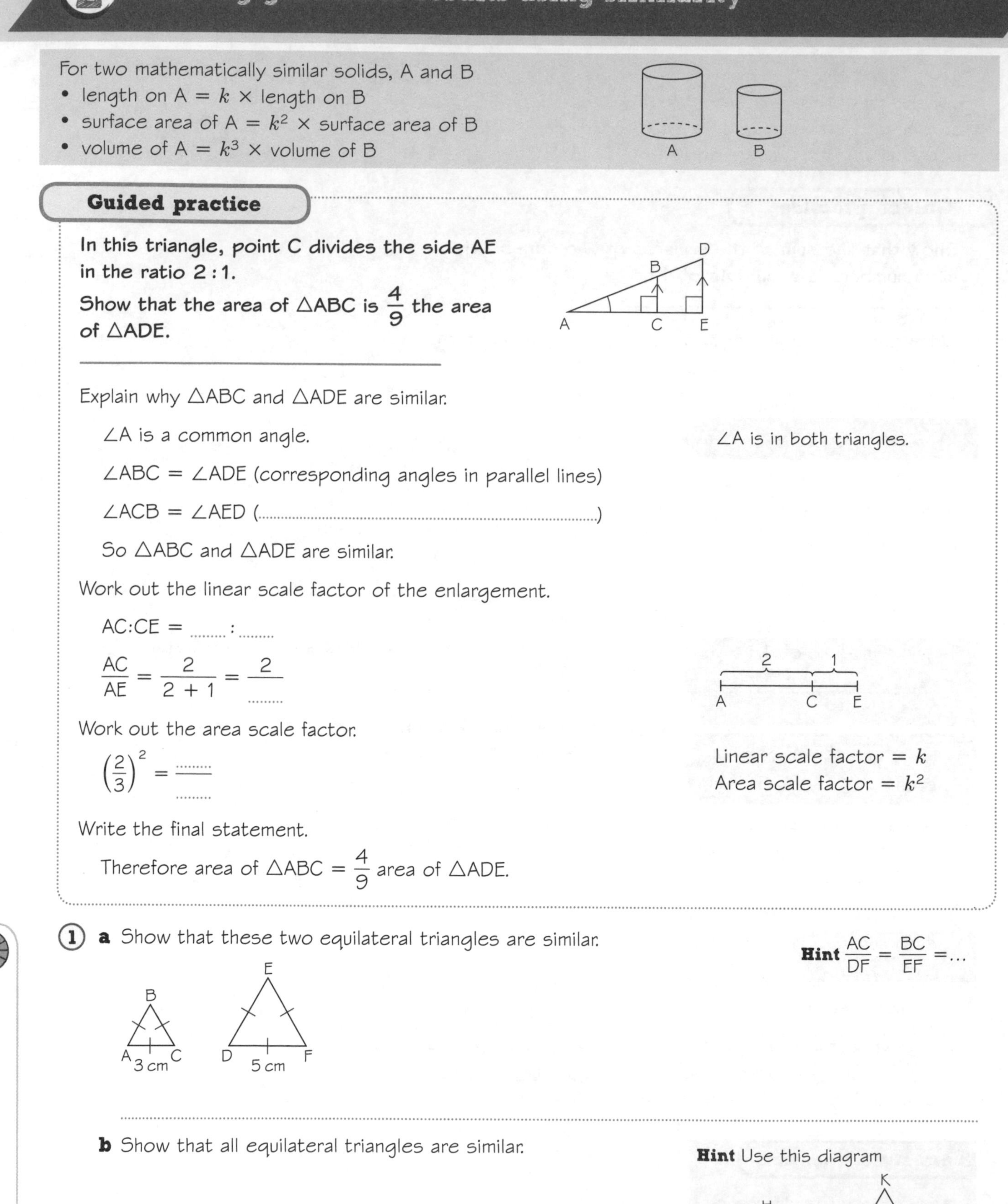

Guided practice

In this triangle, point C divides the side AE in the ratio 2 : 1.
Show that the area of △ABC is $\frac{4}{9}$ the area of △ADE.

Explain why △ABC and △ADE are similar.

∠A is a common angle.

∠A is in both triangles.

∠ABC = ∠ADE (corresponding angles in parallel lines)

∠ACB = ∠AED (..)

So △ABC and △ADE are similar.

Work out the linear scale factor of the enlargement.

AC:CE = ... : ....

$\frac{AC}{AE} = \frac{2}{2+1} = \frac{2}{\dots\dots}$

Work out the area scale factor.

$\left(\frac{2}{3}\right)^2 = \frac{\dots\dots}{\dots\dots}$

Linear scale factor = k
Area scale factor = k^2

Write the final statement.

Therefore area of △ABC = $\frac{4}{9}$ area of △ADE.

1 **a** Show that these two equilateral triangles are similar.

Hint $\frac{AC}{DF} = \frac{BC}{EF} = \dots$

....

b Show that all equilateral triangles are similar.

Hint Use this diagram

and the same steps as in part **a**.

....

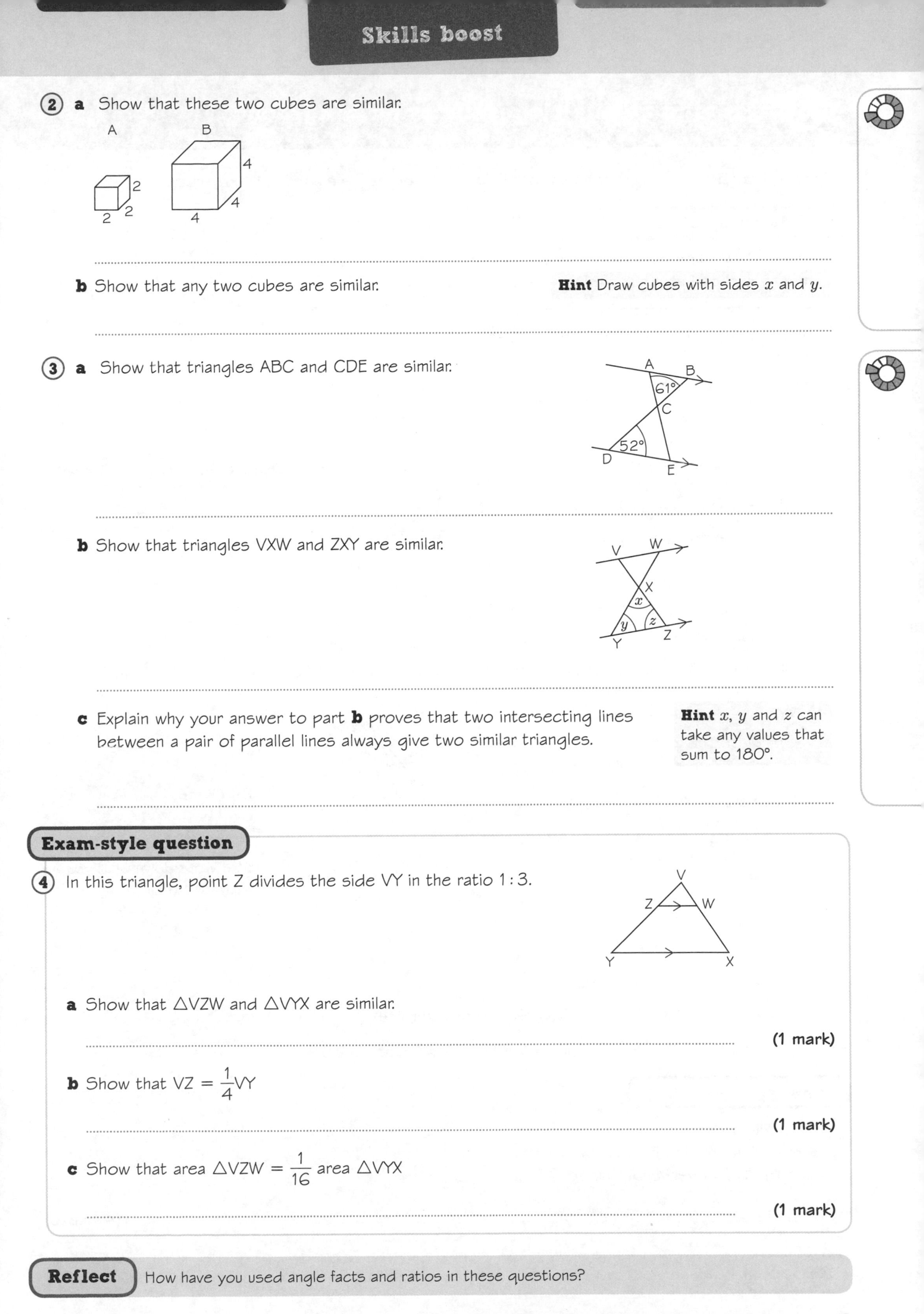

2 **a** Show that these two cubes are similar.

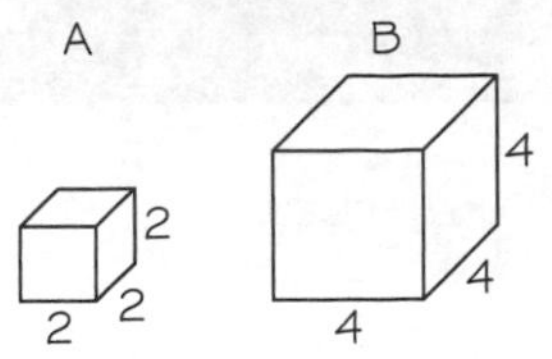

b Show that any two cubes are similar.

Hint Draw cubes with sides x and y.

3 **a** Show that triangles ABC and CDE are similar.

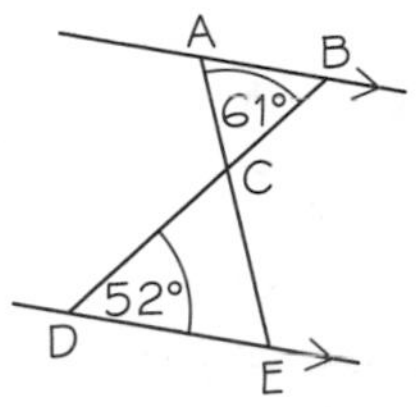

b Show that triangles VXW and ZXY are similar.

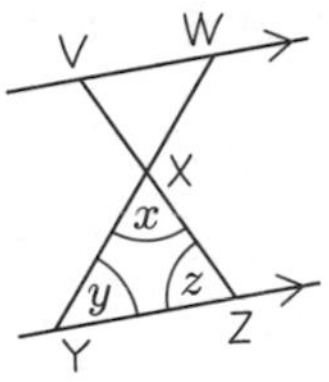

c Explain why your answer to part **b** proves that two intersecting lines between a pair of parallel lines always give two similar triangles.

Hint x, y and z can take any values that sum to 180°.

Exam-style question

4 In this triangle, point Z divides the side VY in the ratio 1 : 3.

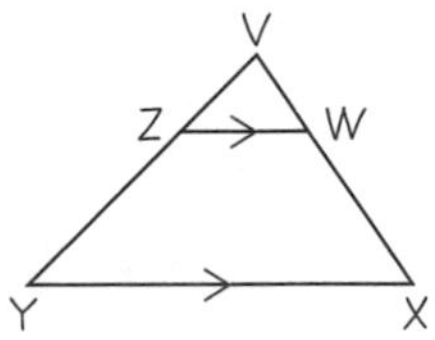

a Show that △VZW and △VYX are similar. **(1 mark)**

b Show that $VZ = \frac{1}{4}VY$ **(1 mark)**

c Show that area $\triangle VZW = \frac{1}{16}$ area $\triangle VYX$ **(1 mark)**

Reflect How have you used angle facts and ratios in these questions?

3 Proving geometric results using congruence

Proving that triangles are congruent can help you to prove other geometric results.

Guided practice

Prove that a line from the centre of a circle that meets a chord at right angles, bisects the chord.

Write down what you know from the question.

OB is a line from the centre.

OB meets chord AC at right angles.

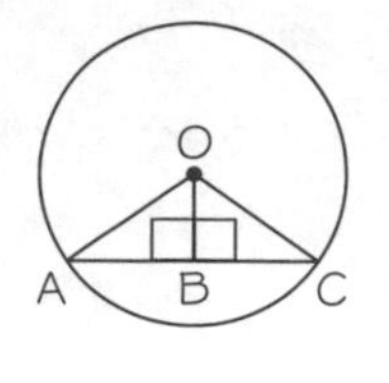

Write down what you need to prove.

Prove that AB = BC

'Bisect' means 'cut in half' so AB = AC

OA = OC (......................)

OB = OB (common side)

Give a reason for each statement.

∠OBA = ∠OBC = ...° (given)

So △OAB and △OCB are congruent (RHS).

Use congruence to explain the statements you need to prove.

Therefore AB = BC, which means that OB bisects the chord AC.

1. XYZ is an isosceles triangle.
   Complete this proof that a line from X that bisects YZ meets YZ at right angles.

   Prove that ∠YWX = ∠........ = ...

   Hint Write down what you need to prove.

   So triangle XYW and triangle XZW are ....

   XW bisects ....

   Hint Write down what you know from the diagram.

   XY = ...

   YW = ...

   Therefore ∠YWX = ∠ZWX

   and ∠YWX + ∠ZWX = ...° (angles on a straight line)

   So ∠YWX = ∠ZWX = ...° and XW meets YZ at right angles.

Exam-style question

2. PQRS is a kite.
   Prove that the line PR that bisects the line QS is perpendicular to QS.

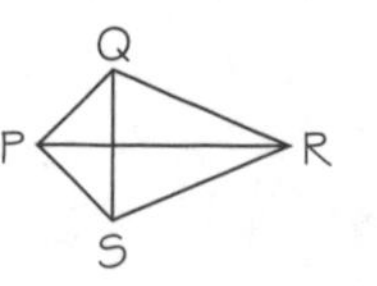

.... **(4 marks)**

Reflect How have you used angle and side properties of triangles and quadrilaterals in these questions?

4 Proving circle theorems

You can use one theorem or angle fact as 'evidence' to prove another.

Guided practice

Prove that the angle at the centre is twice the angle at the circumference.

Prove that $\angle AOC = 2\angle ABC$

Draw a line through OB to make two triangles.

Mark on equal lengths.

$OB = OC =$ .... (radii)

In $\triangle ABO$, label the base angles x.

$\angle OBA = \angle OAB = x$ (base angles in isosceles triangle)

In $\triangle BOC$, label the base angles y.

$\angle OBC = \angle$ .... $= y$ (base angles in isosceles triangle)

$\angle DOC = y + y = 2y$ (exterior angle = sum of interior opposite angles)

$\angle DOA =$ .... (exterior angle = sum of interior opposite angles)

So $\angle AOC = 2x + 2y = 2(x + y)$

$\angle ABC =$ .... + ...

Therefore $\angle AOC = 2\angle ABC$

1. Complete this proof that the angle in a semicircle is a right angle.

   $\angle XOZ =$ ......° (angles on a ....)

   $\angle XOZ = 2\angle$...... (angle at the centre is ....)

   Therefore $180° = 2\angle XYZ$

   $\angle XYZ =$ ....

2. P, Q, R and S are points on the circumference of a circle, centre O.

   a Draw in the lines RO and RS.

   b Find $\angle ROS$. Give a reason. ....

   c Use $\angle ROS$ to find $\angle RQS$. Give a reason. ....

Exam-style question

3. Prove that angles in the same segment are equal.

   .... **(4 marks)**

Reflect In the guided practice, you proved that the angle at the centre is twice the angle at circumference. Which other circle theorems can you use this to prove?

Practise the methods

Answer this question to check where to start.

Check up

Tick the correct proof that a diagonal divides a rectangle into two congruent triangles.

A B
D C

A
AB = DC
AD = BC
BD is common.
So △ADB is congruent to △CBD.

B
Let AB = 4 cm and BC = 3 cm
AB = DC = 4 cm
BC = AD = 3 cm
DB = $\sqrt{3^2 + 4^2}$ = 5 cm
△ADB and △CBD have sides 3, 4 and 5 cm and so are congruent

C
BD is a line of symmetry, so △ADB = △CBD

D
AD = BC
AB = DC
(opposite sides of rectangle are equal)
BD is common.
△ADB is congruent to △CBD (SSS).

If you ticked D go to Q2

If you ticked A, B or C go to Q1 for more practice.

1. Prove that a diagonal divides a parallelogram into two congruent triangles.
Give a reason for every statement you make.

D E
G F

Exam-style question

2. Prove that the difference between an odd and an even number is always odd.
(3 marks)

3. ABCD and AEFG are rectangles.
The point E divides AB in the ratio 2:1.
The point G divides AD in the ratio 2:1.
Show that rectangles AEFG and ABCD are similar.

A E B
F
G
D C

(3 marks)

4. The point P is outside rectangle WXYZ.
PX = PW
Prove that PZ = PY.

P
W X
Z Y

(4 marks)

5. Prove that opposite angles in a cyclic quadrilateral sum to 180°.
(4 marks)

Problem-solve!

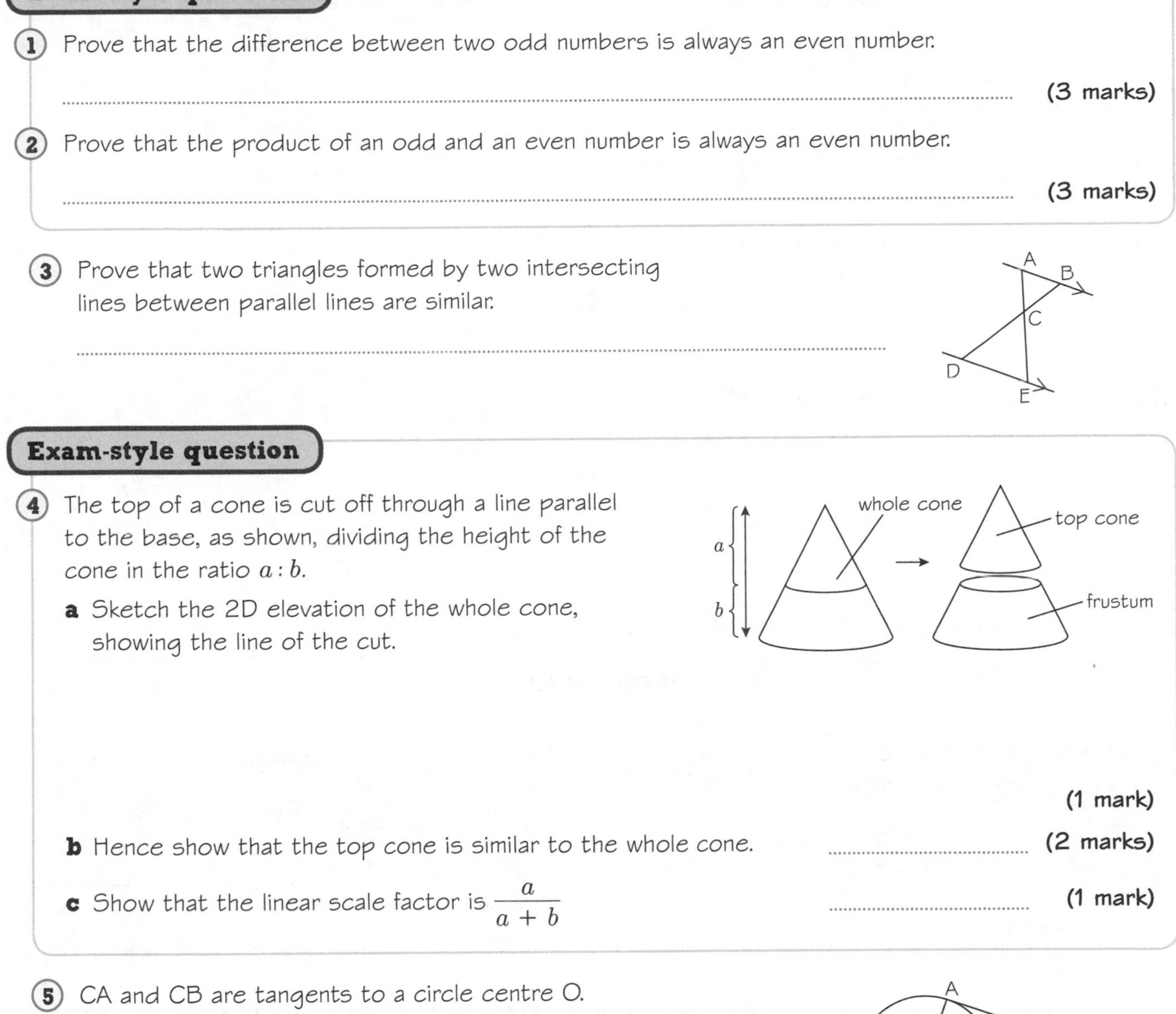

Exam-style questions

1. Prove that the difference between two odd numbers is always an even number.

   .... **(3 marks)**

2. Prove that the product of an odd and an even number is always an even number.

   .... **(3 marks)**

3. Prove that two triangles formed by two intersecting lines between parallel lines are similar.

   ....

Exam-style question

4. The top of a cone is cut off through a line parallel to the base, as shown, dividing the height of the cone in the ratio $a : b$.

   a Sketch the 2D elevation of the whole cone, showing the line of the cut.

   (1 mark)

   b Hence show that the top cone is similar to the whole cone. .... **(2 marks)**

   c Show that the linear scale factor is $\frac{a}{a + b}$ .... **(1 mark)**

5. CA and CB are tangents to a circle centre O.

   a Write down the size of ∠OAC and ∠OBC. Give reasons.

   ....

   b Prove that △OAC is congruent to △OBC.

   ....

   c Hence prove that tangents from a point to a circle are equal.

   ....

A
O
C
B

Now that you have completed this unit, how confident do you feel?

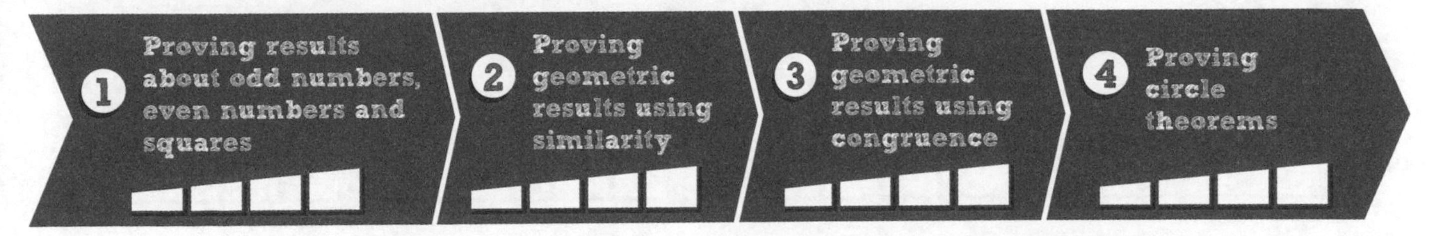

Get started

9 Circles

This unit will help you to solve problems involving graphs, tangents and segments of circles.

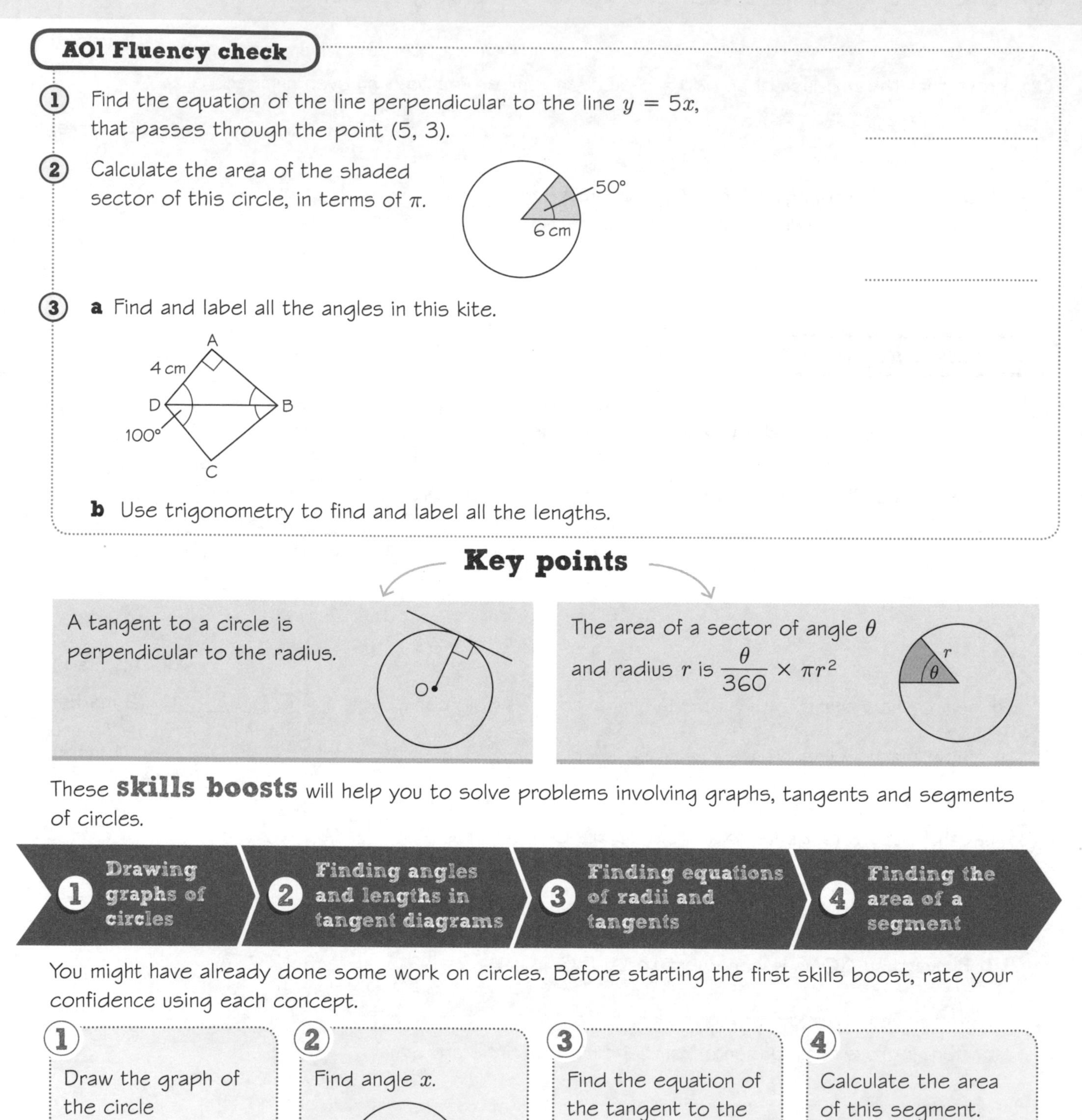

AO1 Fluency check

1. Find the equation of the line perpendicular to the line $y = 5x$, that passes through the point (5, 3).

2. Calculate the area of the shaded sector of this circle, in terms of π.

3. **a** Find and label all the angles in this kite.

   b Use trigonometry to find and label all the lengths.

Key points

A tangent to a circle is perpendicular to the radius.

The area of a sector of angle θ and radius r is $\frac{\theta}{360} \times \pi r^2$

These **skills boosts** will help you to solve problems involving graphs, tangents and segments of circles.

1. Drawing graphs of circles
2. Finding angles and lengths in tangent diagrams
3. Finding equations of radii and tangents
4. Finding the area of a segment

You might have already done some work on circles. Before starting the first skills boost, rate your confidence using each concept.

1. Draw the graph of the circle $x^2 + y^2 = 4$

2. Find angle x.

3. Find the equation of the tangent to the circle $x^2 + y^2 = 25$ at the point (4, −3).

4. Calculate the area of this segment.

How confident are you?

1 Drawing graphs of circles

The graph of $x^2 + y^2 = r^2$ is a circle, centre the origin, radius r.

Guided practice

Draw the graph of $x^2 + y^2 = 81$

Write down r^2 from the equation. Find r.

$x^2 + y^2 = r^2$ — This is the equation of a circle, centre (0, 0).

$x^2 + y^2 = 81$

$r^2 =$ ....

$r = 9$ — So the circle has radius 9 and centre (0, 0).

Mark where the graph crosses the axes.
When $x = 0$, $y^2 = 81$, $y = \pm 9$
When $y = 0$, $x^2 = 81$, $x = \pm 9$

Use compasses to draw a circle, centre O, through the points.

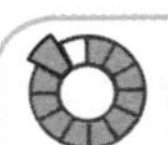

1 **a** Draw the graph of $x^2 + y^2 = 16$

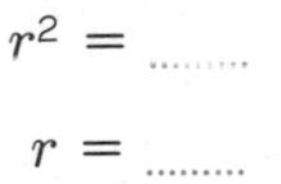

$r^2 =$ ....

$r =$ ....

Circle, radius ...., centre ....

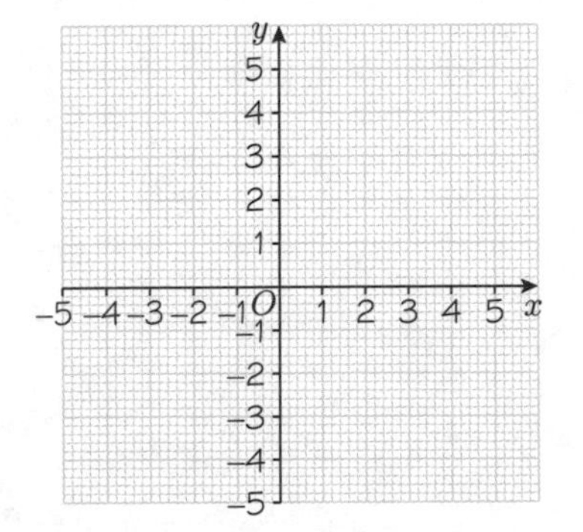

b Draw the graph of $x^2 + y^2 = 9$

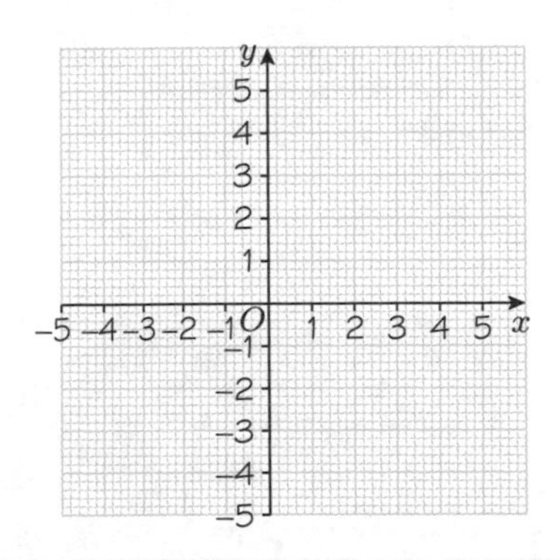

2 Draw the graph of

Hint Round r to 1 decimal place. Label the intercepts on the axes.

a $x^2 + y^2 = 20.25$

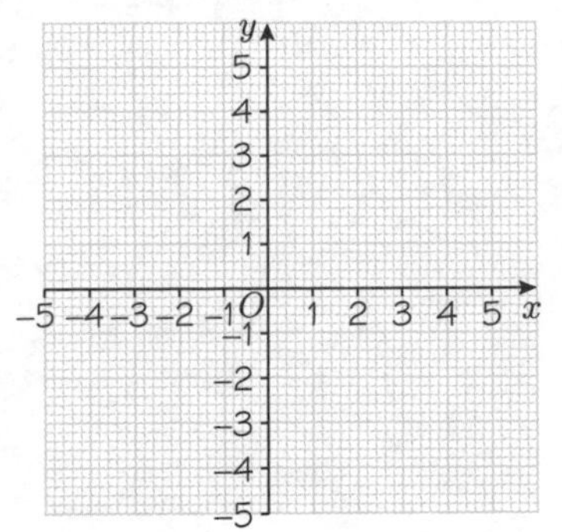

b $x^2 + y^2 = 10$

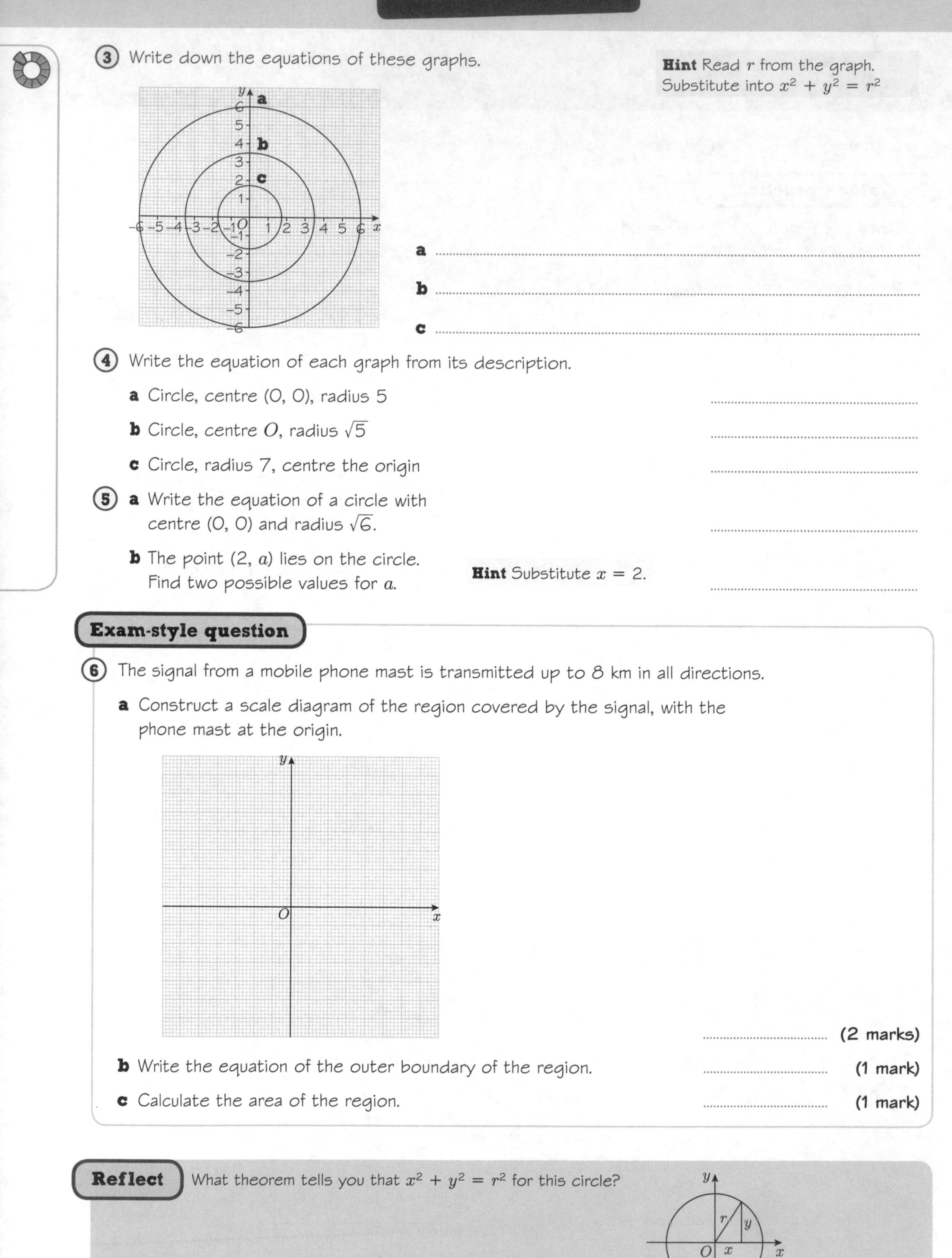

③ Write down the equations of these graphs.

Hint Read r from the graph. Substitute into $x^2 + y^2 = r^2$

a ....

b ....

c ....

④ Write the equation of each graph from its description.

a Circle, centre (0, 0), radius 5 ....

b Circle, centre O, radius $\sqrt{5}$ ....

c Circle, radius 7, centre the origin ....

⑤ **a** Write the equation of a circle with centre (0, 0) and radius $\sqrt{6}$. ....

b The point $(2, a)$ lies on the circle. Find two possible values for a. ....

Hint Substitute $x = 2$.

Exam-style question

⑥ The signal from a mobile phone mast is transmitted up to 8 km in all directions.

a Construct a scale diagram of the region covered by the signal, with the phone mast at the origin.

.... **(2 marks)**

b Write the equation of the outer boundary of the region. .... **(1 mark)**

c Calculate the area of the region. .... **(1 mark)**

Reflect What theorem tells you that $x^2 + y^2 = r^2$ for this circle?

2 Finding angles and lengths in tangent diagrams

Tangents from a point to a circle are equal.

To prove this theorem, see Unit 8, Problem-solve! Q5.

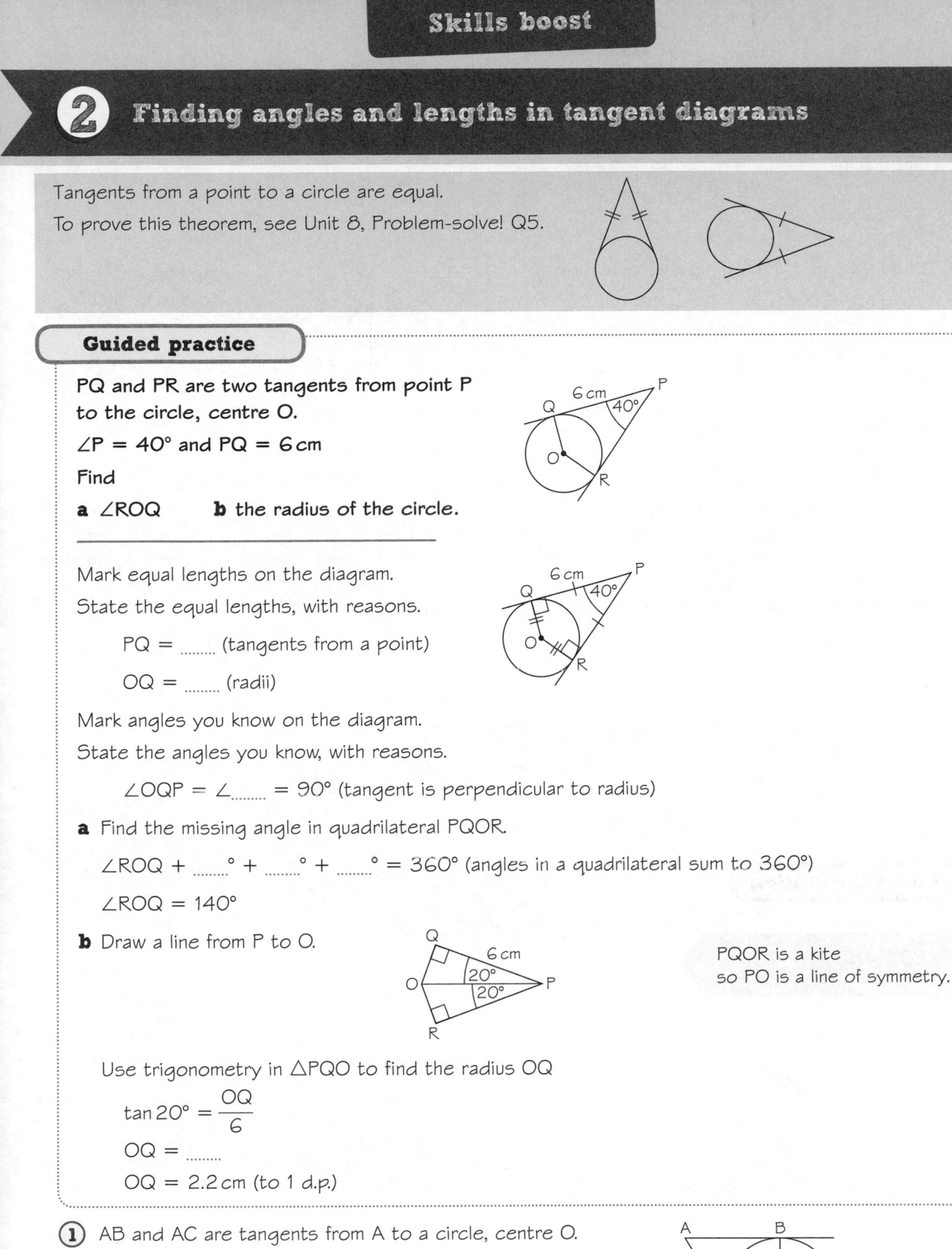

Guided practice

PQ and PR are two tangents from point P to the circle, centre O.

$\angle P = 40°$ and PQ = 6 cm

Find

a $\angle ROQ$ **b** the radius of the circle.

Mark equal lengths on the diagram.

State the equal lengths, with reasons.

PQ = ... (tangents from a point)

OQ = ... (radii)

Mark angles you know on the diagram.

State the angles you know, with reasons.

$\angle OQP = \angle$........ $= 90°$ (tangent is perpendicular to radius)

a Find the missing angle in quadrilateral PQOR.

$\angle ROQ +$ ....° + ...° + ...° = 360° (angles in a quadrilateral sum to 360°)

$\angle ROQ = 140°$

b Draw a line from P to O.

PQOR is a kite so PO is a line of symmetry.

Use trigonometry in $\triangle PQO$ to find the radius OQ

$\tan 20° = \dfrac{OQ}{6}$

OQ = ...

OQ = 2.2 cm (to 1 d.p.)

1 AB and AC are tangents from A to a circle, centre O.

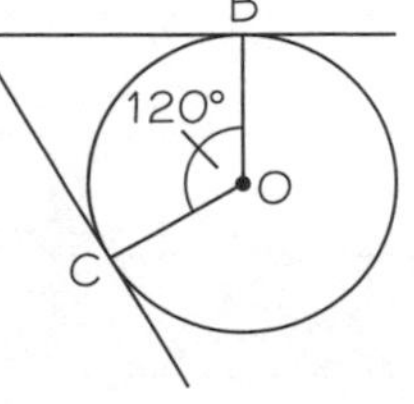

a Find $\angle OBA$ and $\angle OCA$. Give reasons.

....

b Find $\angle BAC$. Give reasons.

....

2 PS is a tangent to a circle, centre O, radius 4 cm.
PO = 10 cm
Find PS.

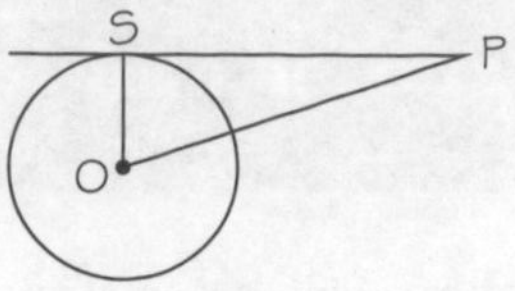

Hint State the angle you know, with a reason.

....

3 XY and XZ are tangents to a circle centre O.
$\angle X = 50°$

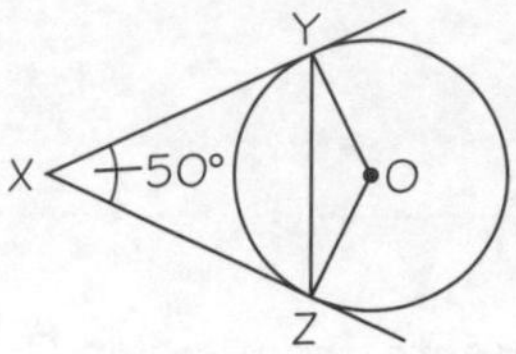

Hint What type of triangle is $\triangle OYZ$?

Find

a $\angle YOZ$ ....

b $\angle OYZ$ ....

Give reasons for your answers.

....

4 MN and MP are tangents to a circle centre O, radius 5 cm.
$\angle NOP = 110°$.

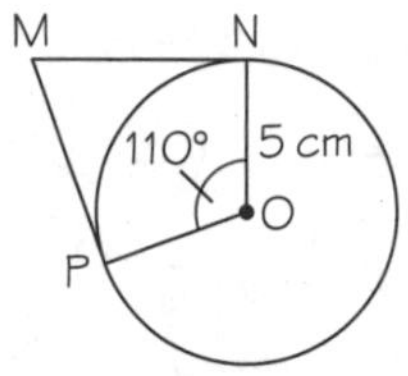

a Find angle NMP. ....

b Explain why the line MO bisects angles NMP and NOP.

....

c Use triangle MNO to find the length of MO. ....

Exam-style question

5 DE and DF are tangents to a circle, centre O, radius 7 cm.
$\angle EOF = 150°$

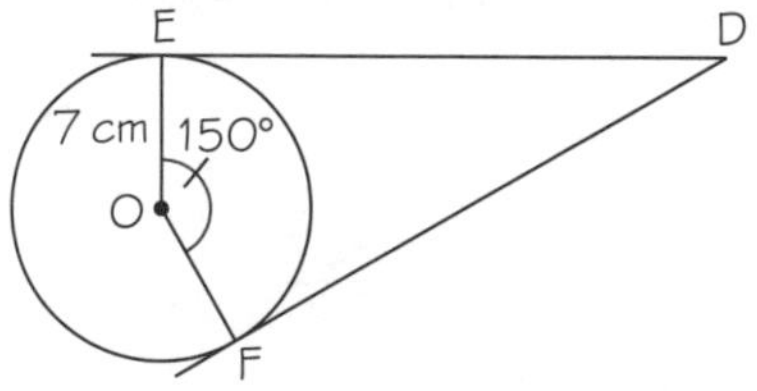

Find the length OD.
Give reasons for each stage of your working.

....

....

....

.... **(4 marks)**

Reflect How have you used properties of right-angled triangles and kites in these questions?

3 Finding equations of radii and tangents

The tangent is perpendicular to the radius.
When a line has gradient m, the gradient of a line perpendicular to it is $-\frac{1}{m}$

Guided practice

The equation of this circle is $x^2 + y^2 = 169$
Find the equation of
a the radius
b the tangent to the circle at the point A $(-5, 12)$.

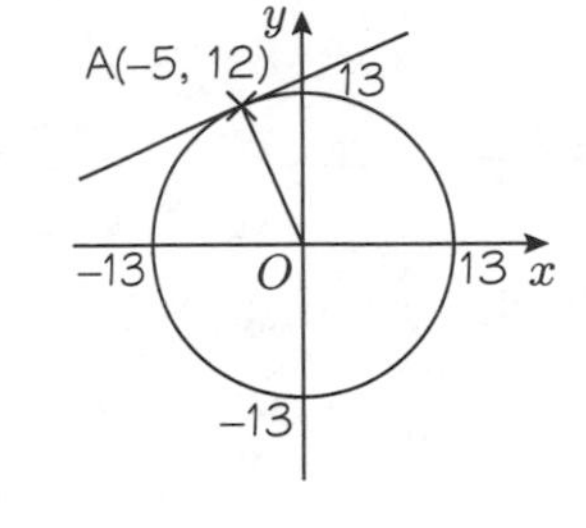

a Find the gradient of the radius, OA.

Gradient of radius $= \frac{12}{\dots\dots} = -\frac{12}{5}$

Gradient $= \frac{\text{change in } y}{\text{change in } x}$

Use $y = mx + c$

$y = \dots\dots x + c$

In $y = mx + c$, c is the y-intercept.

y-intercept is 0, so $c = 0$.

Equation of radius OA is $y = -\frac{12}{5}x$

b Find the gradient of the tangent at A.

Gradient of tangent is $\frac{5}{12}$

Tangent and radius are perpendicular.
Gradient of perpendicular $= -\frac{1}{m}$

Use $y = mx + c$

$y = \frac{5}{12}x + c$

Substitute $(-5, 12)$ to find c.

$12 = \frac{5}{12} \times \dots\dots + c$

$c = \dots\dots + \frac{25}{12} = \frac{169}{12}$

Equation of tangent is

$y = \frac{5}{12}x + \frac{169}{12}$

For an equation in the form $ax + by = c$, multiply both sides by 12 and rearrange:
$12y - 5x = 169$

1 For this circle, find
i the gradient of the radius
ii the equation of the radius
at point A (12, 5)

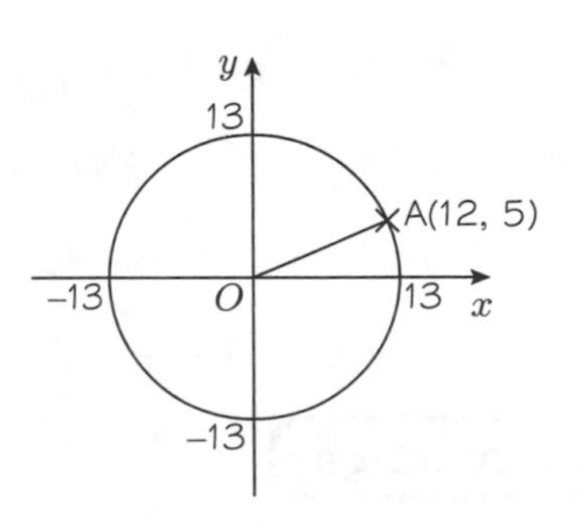

a Gradient of OA $= \frac{\dots\dots}{\dots\dots}$

b Equation of OA: $y = mx + c$

$y = \dots\dots x$

② For this circle, find

i the gradient of the radius

ii the equation of the radius

at each point.

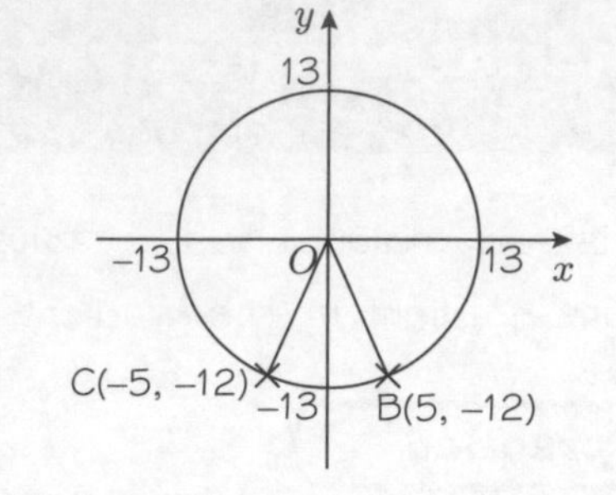

a at B (5, −12)

i Gradient of OB = $\frac{\dots\dots}{\dots\dots}$

ii Equation of OB: $y = mx + c$

$y = \dots\dots x$

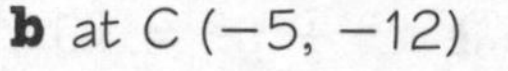

b at C (−5, −12)

i Gradient of OC = $\frac{\dots\dots}{\dots\dots}$

ii Equation of OC: $y = mx + c$

$y = \dots\dots x$

③ This circle has equation $x^2 + y^2 = 25$.

P is the point (−3, −4).

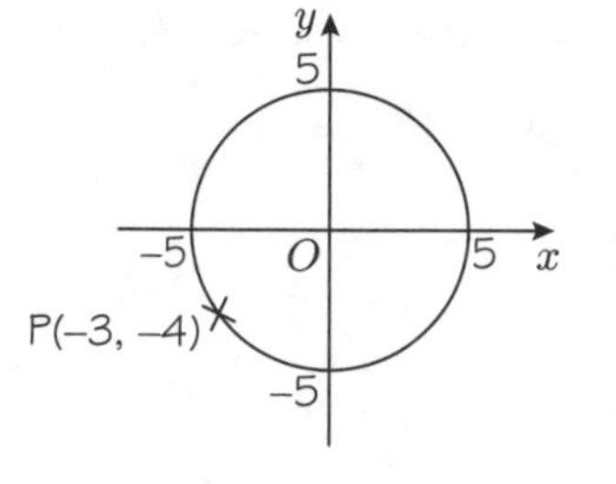

a Find the equation of the radius at P. ....

b Find the gradient of the tangent at P. ....

Hint $-\frac{1}{m}$

c Find the equation of the tangent at P. ....

Hint Substitute (−3, −4).

④ Find the equation of

a the radius ....

b the tangent ....

to the circle $x^2 + y^2 = 100$ at point M (6, 8).

Hint Sketch the graph and mark the point M.

⑤ Find the equation of the tangent to the circle $x^2 + y^2 = 225$ at the point Q (−9, 12).

Give your answer in the form $ax + by = c$. ....

⑥ PT and ST are tangents to the circle $x^2 + y^2 = 100$ at the points P(−6, 8) and S(−8, −6) respectively.

a Find the equations of the radii OP and OS. ....

b Hence show that OP and OS are perpendicular. ....

c Find the equations of tangents PT and ST. ....

d Hence find the coordinates of T. ....

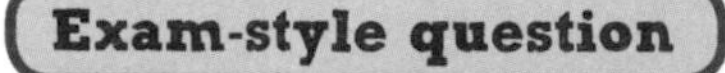

Exam-style question

⑦ Find the equation of the tangent to the circle $x^2 + y^2 = 289$ at the point T (15, 8). .... **(4 marks)**

Reflect How have you used the tangent theorem in these questions?

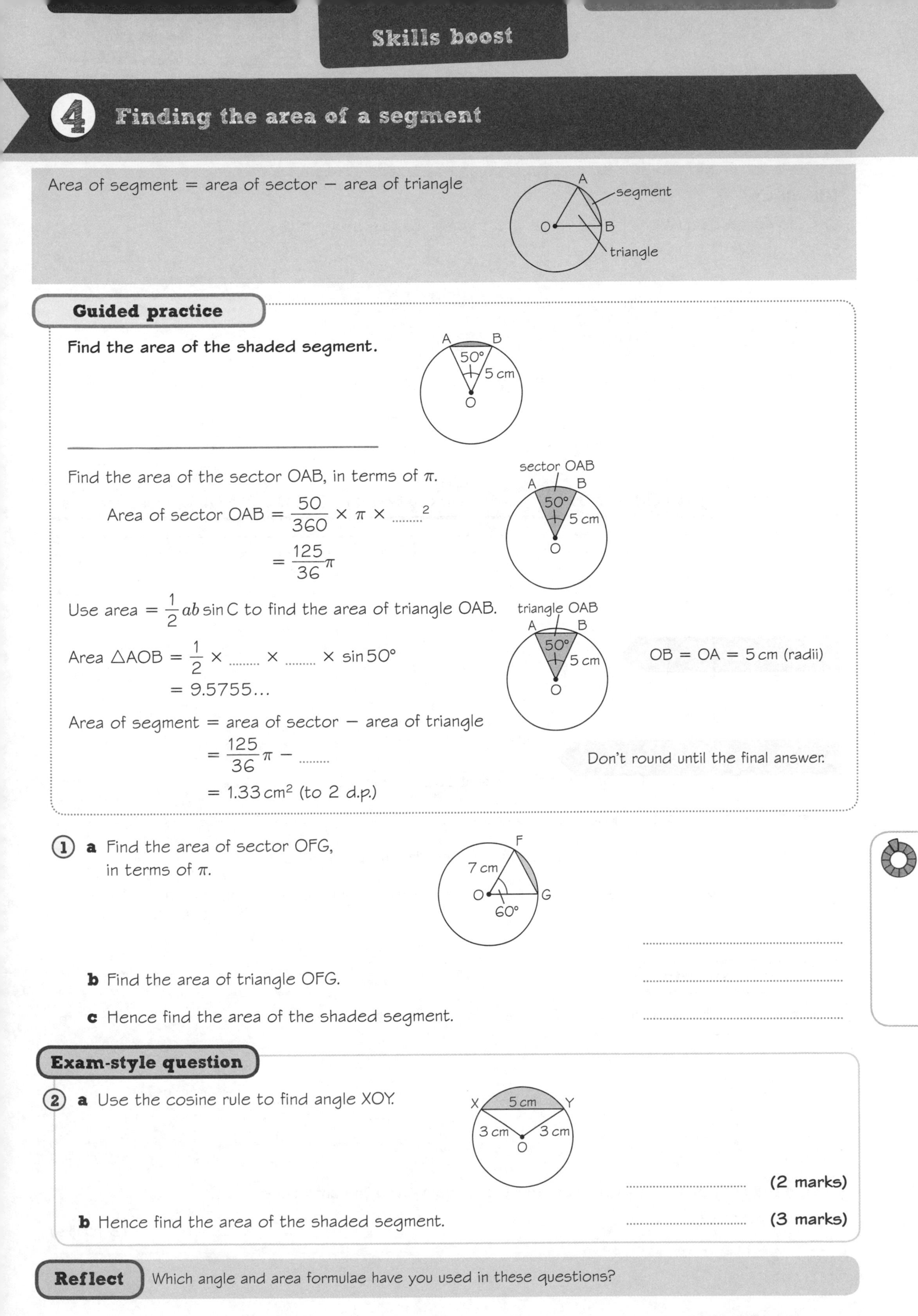

4 Finding the area of a segment

Area of segment = area of sector − area of triangle

Guided practice

Find the area of the shaded segment.

Find the area of the sector OAB, in terms of π.

$$\text{Area of sector OAB} = \frac{50}{360} \times \pi \times \dots\dots^2$$
$$= \frac{125}{36}\pi$$

Use area $= \frac{1}{2}ab\sin C$ to find the area of triangle OAB.

$$\text{Area } \triangle AOB = \frac{1}{2} \times \dots\dots \times \dots\dots \times \sin 50°$$
$$= 9.5755\dots$$

OB = OA = 5 cm (radii)

Area of segment = area of sector − area of triangle

$$= \frac{125}{36}\pi - \dots\dots$$
$$= 1.33\text{ cm}^2 \text{ (to 2 d.p.)}$$

Don't round until the final answer.

1. **a** Find the area of sector OFG, in terms of π. ....

   b Find the area of triangle OFG. ....

   c Hence find the area of the shaded segment. ....

Exam-style question

2. **a** Use the cosine rule to find angle XOY. .... **(2 marks)**

   b Hence find the area of the shaded segment. .... **(3 marks)**

Reflect Which angle and area formulae have you used in these questions?

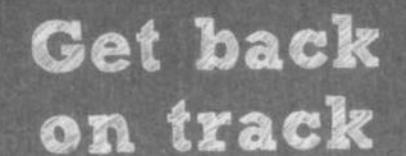

Practise the methods

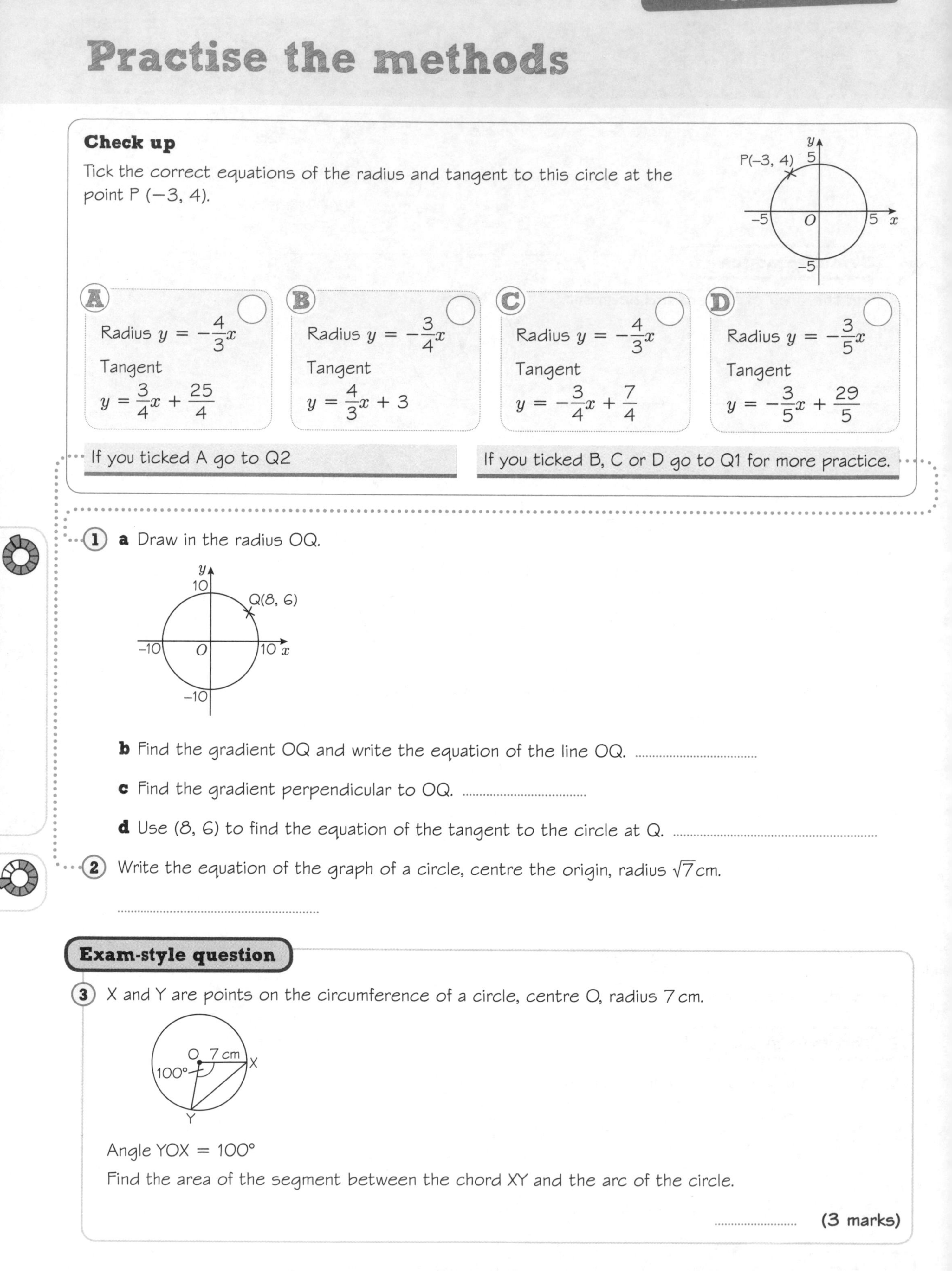

Check up

Tick the correct equations of the radius and tangent to this circle at the point P (−3, 4).

A Radius $y = -\frac{4}{3}x$ Tangent $y = \frac{3}{4}x + \frac{25}{4}$

B Radius $y = -\frac{3}{4}x$ Tangent $y = \frac{4}{3}x + 3$

C Radius $y = -\frac{4}{3}x$ Tangent $y = -\frac{3}{4}x + \frac{7}{4}$

D Radius $y = -\frac{3}{5}x$ Tangent $y = -\frac{3}{5}x + \frac{29}{5}$

If you ticked A go to Q2

If you ticked B, C or D go to Q1 for more practice.

1 **a** Draw in the radius OQ.

b Find the gradient OQ and write the equation of the line OQ. ....

c Find the gradient perpendicular to OQ. ....

d Use (8, 6) to find the equation of the tangent to the circle at Q. ....

2 Write the equation of the graph of a circle, centre the origin, radius $\sqrt{7}$ cm.

....

Exam-style question

3 X and Y are points on the circumference of a circle, centre O, radius 7 cm.

Angle YOX = 100°

Find the area of the segment between the chord XY and the arc of the circle.

.... **(3 marks)**

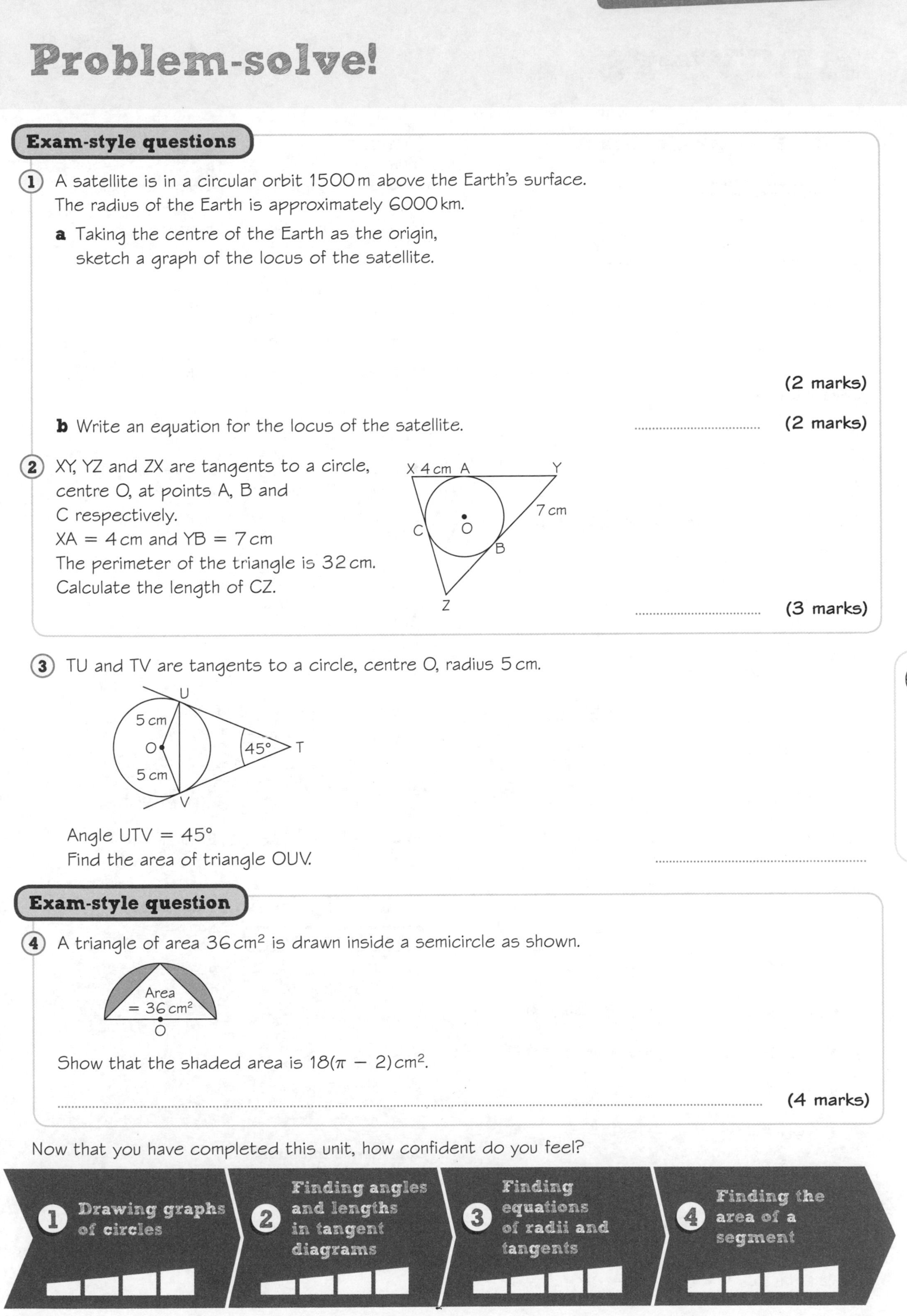

Problem-solve!

Exam-style questions

1 A satellite is in a circular orbit 1500 m above the Earth's surface.
The radius of the Earth is approximately 6000 km.

a Taking the centre of the Earth as the origin, sketch a graph of the locus of the satellite.

(2 marks)

b Write an equation for the locus of the satellite. .... **(2 marks)**

2 XY, YZ and ZX are tangents to a circle, centre O, at points A, B and C respectively.
XA = 4 cm and YB = 7 cm
The perimeter of the triangle is 32 cm.
Calculate the length of CZ.

.... **(3 marks)**

3 TU and TV are tangents to a circle, centre O, radius 5 cm.

Angle UTV = 45°
Find the area of triangle OUV. ....

Exam-style question

4 A triangle of area 36 cm^2 is drawn inside a semicircle as shown.

Show that the shaded area is $18(\pi - 2)$ cm^2.

.... **(4 marks)**

Now that you have completed this unit, how confident do you feel?

1 Drawing graphs of circles

2 Finding angles and lengths in tangent diagrams

3 Finding equations of radii and tangents

4 Finding the area of a segment

Answers

Unit 1 Further algebra

AO1 Fluency check

1. **a** $\frac{m^2}{4n^2}$ **b** $\frac{2\sqrt{y}}{3}$ **c** $\frac{8t^3}{x^3}$ **d** $2\sqrt[3]{\frac{y}{a}}$
2. **a** $x = \frac{y}{2z}$ **b** $x = \frac{4t}{3y}$ **c** $x = \frac{\sqrt{y}}{2}$
   d $x = t^2$ **e** $x = \sqrt{\frac{2V}{m}}$ **f** $x = k^2 - y$
3. **a** $x^2 - 4x - 5$ **b** $x^2 + 2x - 8$
   c $2x^2 - 9x - 5$ **d** $8x^2 + 2x - 3$

Confidence check

1. $t = \frac{x}{36y^2}$
2. $8x^2 - 10ax - 3a^2$
3. $x^3 + 6x^2 + 11x + 6$

Skills boost 1 Rearranging complex formulae

Guided practice

$abx + \underline{ac} = 4x + e$
$abx - \underline{4x} = e - ac$
$x(\underline{ab} - 4) = e - ac$

1. **a** $x = \frac{at}{y + 5}$ **b** $m = \frac{E}{gh + \frac{1}{2}v^2}$
2. **a** $t = \frac{g - fr}{r - s}$ **b** $t = \frac{2b + 3c}{3 - 2a}$
   c $t = \frac{b - 3ab}{3a + 2}$ **d** $t = \frac{x + 2ax}{4b + 3}$
3. **a** $y = \frac{3}{x - 1}$ **b** $y = \frac{-3n - t}{nr - x} = \frac{3n + t}{x - nr}$
4. **a** $x = \frac{am^2}{r^2}$ **b** $x = \frac{t}{36y^2}$
   c $x = \sqrt{\frac{m}{y}} - 3$ **d** $x = \frac{a\sqrt{2y}}{3}$
5. **a** $y = \sqrt[3]{\frac{t}{a}}$ **b** $y = 2\sqrt[3]{z}$
   c $y = b\sqrt[3]{rs}$ **d** $y = \sqrt[3]{\frac{p}{n}} + 5$
   e $y = \sqrt[3]{tx} - 2$ **f** $y = \sqrt[3]{2b}(a - 1)$
6. **a** $x = \frac{m^3}{4}$ **b** $x = vy^3$
   c $x = \left(\frac{r}{at}\right)^3$ **d** $x = \frac{by^3}{a}$
   e $x = yd^3$ **f** $x = \frac{b}{e^3}$
7. $m = \frac{5t^2}{2 + t^2}$

Skills boost 2 Expanding two or three brackets

Guided practice

$(x + 2)(x^2 + 3x - 1) = x^3 + \underline{3}x^2 - x + 2x^2 + \underline{6}x - 2$

1. **a** $6x^2 - ax - a^2$ **b** $8m^2 + 6my - 9y^2$
2. **a** $3x^2 + 6x + 3$ **b** $-2x^2 - 2x + 10$
   c $x^3 + x^2 + 2x$ **d** $x^3 + 3x^2 - x$
3. **a** $x^3 + 6x^2 + 9x + 2$ **b** $x^3 - 4x^2 + x + 6$
4. **a** $x^3 + 7x^2 + 14x + 8$ **b** $x^3 + 4x^2 + x - 6$
5. **a** $x^3 + 6x^2 + 5x$ **b** $x^3 + 6x^2 + 12x + 8$

Practise the methods

1. **a** $y = 18x^2$ **b** $y = ac^2d^2$ **c** $y = 12p^2q^2r$
2. **a** $z = \frac{t}{w^2}$ **b** $z = \frac{m}{v^2}$ **c** $z = \frac{n}{9s^2}$
3. **a** $2a^2 + 7ab + 3b^2$ **b** $10s^2 - st - 3t^2$
4. $x = \frac{4 - dt}{at - b}$
5. $n = \frac{s}{s + 2}$
6. **a** $x^3 - 7x + 6$ **b** $x^3 - 5x^2 - 4x + 20$
7. **a** $x^3 - x^2 - 16x + 16$ **b** $x^3 + 3x^2 + 3x + 1$
8. $r = \sqrt[3]{\frac{3V}{4\pi}}$

Problem-solve!

1. $(x + 3)^3 = x^3 + 9x^2 + 27x + 27$
2. $t = \frac{3z + 7 - 3b}{b + 3a}$
3. $x^3 + 5x^2 + 3x$
4. $y = \sqrt{\frac{80 - m^2}{32}}$
5. **a** $a^2 = b^2 + c^2 - 2bc \cos A$
   b $\cos A = \frac{b^2 + c^2 - a^2}{2bc}$
6. **a** $x = \sqrt{\frac{z + 3y}{a - 1}}$ **b** $x = \sqrt{\frac{qn - pm}{2p + 3q}}$
7. $x = s^3y - 2$
8. **a** $2x^3 + 3x^2 - 3x - 2$ **b** $3x^3 + 4x^2 - x - 2$
9. **a** -1 **b** $27 - 10\sqrt{2}$

Unit 2 Algebraic fractions

AO1 Fluency check

1. **a** $x(2x - 5)$ **b** $(x - 5)(x + 5)$
   c $(x + 1)(x + 5)$ **d** $(2x + 3)(x - 2)$
2. **a** $\frac{1}{x}$ **b** $6x^2$ **c** $\frac{1}{s}$ **d** $\frac{4xy}{3}$
3. **a** $\frac{17x}{20}$ **b** $\frac{1}{3x}$ **c** $\frac{5x + 2}{6}$
4. **a** $x = -5$ or $x = 3$
   b $x = \frac{5 \pm \sqrt{17}}{4}$ or $x = 2.28$, $x = 0.22$ (to 2 d.p.)

Confidence questions

1. $\frac{x - 1}{x - 3}$
2. $\frac{2x - 3}{(x + 1)(x - 4)}$
3. $x = 0$ or $x = 3$

Skills boost 1 Simplifying algebraic fractions

Guided practice

$\frac{(x-3)(x+1)}{(x+3)(x+1)}$

1 a $2(x+2)$ b $x-3$ c x

2 a $\frac{x-4}{x+1}$ b $\frac{x+3}{x+2}$

3 a $\frac{x-4}{x+1}$ b $\frac{x-5}{x}$

4 a $\frac{-1}{x-4}$ b $\frac{-4}{x+5}$

5 $\frac{2x+1}{x+1}$

Skills boost 2 Adding and subtracting algebraic fractions

Guided practice

$\frac{x+1}{3(x+2)(x+1)} + \frac{3x+6}{3(x+2)(x+1)}$

1 a $\frac{2x+4}{(x+1)(x+3)}$ or $\frac{2(x+2)}{(x+1)(x+3)}$

b $\frac{2x-2}{(x-4)(x+2)}$ or $\frac{2(x-1)}{(x-4)(x+2)}$

c $\frac{2}{(x-3)(x-1)}$

2 a $\frac{3x+7}{(x+5)(x+1)}$ b $\frac{x+11}{(x-1)(x+2)}$

3 a $\frac{4x-8}{(x-3)(x+1)}$ b $\frac{5x+13}{(x+2)(x+5)}$

c $\frac{27x+14}{(x+2)(3x+1)}$ d $\frac{-3x-12}{(4x+1)(x-2)}$

4 a $\frac{-3x+1}{(4x+2)(x+3)}$

b $\frac{7x-4}{10(x-2)(x+3)}$

5 a $\frac{x^2-x+5}{(x+5)(x-2)}$

b $\frac{4x^2+8x}{(x-1)(x+3)}$ or $\frac{4x(x+2)}{(x-1)(x+3)}$

6 a $\frac{x-4}{(x+3)(2x-1)}$ b $\frac{4}{(2x+1)(2x+5)}$

c $\frac{x^2-3x-6}{(x+3)(x-1)}$ d $\frac{3x^2-5x}{(x^2-1)}$

7 $\frac{3x+7}{(x-1)(2x+3)}$

Skills boost 3 Solving equations involving algebraic fractions

Guided practice

$(x-3)(x+1)$
$0=(x-1)(x-5)$

1 a $x=0$ or $x=3$ b $x=4$ or $x=\frac{-5}{3}$

2 a $x=2$ or $x=-3$ b $x=0.80$ or $x=-1.55$

3 $x=1$ or $x=2$

Practise the methods

1 a $\frac{x-5}{(x+1)(2x-1)}$ b $\frac{-5x+13}{(3x-4)(x+5)}$

2 a x b $\frac{x}{x-1}$ c $\frac{x-7}{x-1}$ d $\frac{x}{x-2}$

3 $\frac{-1}{x+5}$

4 a $\frac{-5}{(x+2)(x-3)}$ b $\frac{17x+28}{3(x+2)(x-1)}$

5 $\frac{2x+3}{x+1}$

6 $\frac{7x-13}{2x^2-9x-5}$

7 $x=0.48$ or $x=-6.73$

Problem-solve!

1 $\frac{x}{x+3}$

2 a $\frac{x+5}{2}$ b cannot simplify

c $\frac{x-3}{y}$ d cannot simplify e $\frac{2x-3}{x}$

3 a $\frac{a+b}{ab}$ b $c=\frac{ab}{a+b}$

4 a $R_T=\frac{R_1R_2}{R_1+R_2}$ b $R_2=\frac{R_TR_1}{R_1-R_T}$

5 $x=\frac{1\pm\sqrt{97}}{8}$

6 $x=-0.44$ or $x=-31.56$

7 $x=-7.5$

Unit 3 Iterative processes

AO1 Fluency check

1 a 2.571 (to 3 d.p.) b −0.40 (to 2 d.p.)

2 a 2.00 b 1.67 c 3.04 d 2.00

3 10, 7, 4.9, 3.43, 2.401

4 32.4

Confidence questions

1 $x_1=2$
$x_2=1$
$x_3=1.414213562$
$x_4=1.259280127$
$x_5=1.319363435$
$x_6=1.296393677$
$x_7=1.30522271$
$x=1.30$ (to 2 d.p.)

2 $x_1=-0.75$
$x_2=-0.927734375$
$x_3=-0.9748115866$
$x_4=-0.9907902684$
$x_5=-0.9965780602$
$x_6=-0.9987211587$
$x=-1.00$ (to 2 d.p.)

Skills boost 1 Solving quadratic equations using an iterative process

Guided practice

No

$x_3=\sqrt{2\times3.410754987+5}$

Yes

1 a $x_1=1.414213562$
$x_2=1.608038071$
$x_3=1.546596886$

b $x_1=2.\dot{3}$
$x_2=1.\dot{1}4\dot{8}$
$x_3=-0.2272519433$

2 $x_1=3.16227766=3.16$ (to 2 d.p.)
$x_2=3.187832753=3.19$ (to 2 d.p.)
$x_3=3.19183846=3.19$ (to 2 d.p.)
$x=3.19$ (to 2 d.p.)

3 $x_1=-0.25$
$x_2=-0.3671875=-0.367$ (to 3 d.p.)
$x_3=-0.3581466675=-0.358$ (to 3 d.p.)
$x_4=-0.3589663706=-0.359$ (to 3 d.p.)
$x_5=-0.3588928931=-0.359$ (to 3 d.p.)
$x=-0.359$ (to 3 d.p.)

④ $x_1 = 4.795\,831\,523 = 4.80$ (to 2 d.p.)
$x_2 = 4.733\,744\,496 = 4.73$ (to 2 d.p.)
$x_3 = 4.746\,842\,214 = 4.75$ (to 2 d.p.)
$x_4 = 4.744\,082\,163 = 4.74$ (to 2 d.p.)
$x_5 = 4.744\,663\,916 = 4.74$ (to 2 d.p.)
$x = 4.74$ (to 2 d.p.)

Skills boost 2 Solving cubic equations using an iterative process

Guided practice

$x_1 = 0$
$x_3 = -0.345\,679\,012 = -0.346$ (to 3 d.p.)
$x_4 = -0.347\,102\,187 = -0.347$ (to 3 d.p.)
$x_5 = -0.347\,272\,949 = -0.347$ (to 3 d.p.)

① **a** $x_1 = 0.9$
$x_2 = 0.172\,9$
$x_3 = 0.100\,516\,874\,3$

b $x_1 = 1$
$x_2 = 1.442\,249\,57$
$x_3 = 1.367\,579\,941$

② $x_1 = 1$
$x_2 = -1.259\,921\,05 = -1.26$ (to 2 d.p.)
$x_3 = -1.122\,034\,712 = -1.12$ (to 2 d.p.)
$x_4 = -1.203\,100\,521 = -1.20$ (to 2 d.p.)
$x_5 = -1.157\,982\,444 = -1.20$ (to 2 d.p.)
$x = -1.20$ (to 2 d.p.)

③ $x_1 = -0.25$
$x_2 = -0.253\,906\,250 = -0.25$ (to 2 d.p.)
$x = -0.25$ (to 2 d.p.)

④ **a** $1^3 + 2 \times 1 = 3 < 5$,
$2^3 + 2 \times 2 = 12 > 5$.
The value of x that gives $x^3 + 2x = 5$ must lie between 1 and 2.

b $x_0 = 1.5$
$x_1 = 1.259\,921\,05$
$x_2 = 1.353\,608\,617$
$x_3 = 1.318\,623\,980$
$x_4 = 1.331\,903\,363$
$x_5 = 1.326\,894\,080$
$x = 1.33$ (to 2 d.p.)

c $1.33^3 + 2 \times 1.33 = 5.012\,637 \approx 5$

⑤ $x_1 = 2.117\,911\,792 = 2.118$ (to 3 d.p.)
$x_2 = 2\,408\,482\,756 = 2.408$ (to 3 d.p.)
$x_3 = 2.551\,035\,251 = 2.551$ (to 3 d.p.)
$x_4 = 2.621\,496\,421 = 2.621$ (to 3 d.p.)
$x_5 = 2.656\,386\,021 = 2.656$ (to 3 d.p.)
$x_6 = 2.673\,670\,618 = 2.674$ (to 3 d.p.)
$x_7 = 2.682\,234\,953 = 2.682$ (to 3 d.p.)
$x_8 = 2.686\,478\,751 = 2.686$ (to 3 d.p.)
$x_9 = 2.688\,581\,691 = 2.689$ (to 3 d.p.)
$x_{10} = 2.689\,623\,779 = 2.690$ (to 3 d.p.)
$x_{11} = 2.690\,140\,176 = 2.690$ (to 3 d.p.)
$x = 2.690$ (to 3 d.p.)

Practise the methods

① **a** $x_1 = -1.709\,975\,947$
$x_2 = -2.229\,089\,043 = -2.230$ (to 3 d.p.)

b $x_1 = 0$
$x_2 = 2.828\,427\,125 = 2.828$ (to 3 d.p.)

② $x_1 = -0.8\dot{3}$
$x_2 = -0.929\,783\,950\,617 = -0.93$ (to 2 d.p.)
$x_3 = -0.967\,299\,424\,5 = -0.97$ (to 2 d.p.)
$x_4 = -0.984\,178\,548\,1 = -0.98$ (to 2 d.p.)
$x_5 = -0.992\,213\,773\,2 = -0.99$ (to 2 d.p.)
$x_6 = -0.996\,137\,120\,6 = -1.00$ (to 2 d.p.)
$x_7 = -0.998\,076\,011\,6 = -1.00$ (to 2 d.p.)
$x = -1.00$ (to 2 d.p.)

③ $x_1 = -0.\dot{3}$
$x_2 = -0.\dot{2}3076\dot{9}$
$x_3 = -0.2\dot{3}\dot{6}$
$x_4 = -0.236\,051\,502\,1$
$x = -0.24$ (to 2 d.p.)

④ $x_1 = 1.817\,120\,593$
$x_2 = 1.885\,022\,855$
$x_3 = 1.861\,139\,4$
$x_4 = 1.869\,709\,9$
$x_5 = 1.866\,656$
$x = 1.87$ (to 2 d.p.)

Problem-solve!

① $x^2 - 3x - 5 = 0$

a i $x(x - 3) = 5$
Divide both sides by $(x - 3)$
$x = \frac{5}{x - 3}$

ii $x^2 - 3x = 5$
Divide both sides by x
$x - 3 = \frac{5}{x}$
Add 3 to both sides
$x = \frac{5}{x} + 3$

b $x_1 = -1.25$
$x_2 = -1.176\,470\,588\,235$
$x_3 = -1.197\,183\,098\,59$
$x_4 = -1.191\,275\,168$
$x_5 = -1.192\,954\,363$
$x = -1.19$ (to 2 d.p.)

c $x_1 = 4.25$
$x_2 = 4.176\,470\,588\,235$
$x_3 = 4.197\,183\,098\,591\,5$
$x_4 = 4.191\,275\,168$
$x_5 = 4.192\,954\,363$
$x = 4.19$ (to 2 d.p.)

d $(-1.19)^2 - 3 \times -1.19 - 5 = -0.0139 \approx 0.01$
$4.19^2 - 3 \times 4.19 - 5 = -0.0139 \approx 0.01$

② **a** $0^3 - 3 \times 0 + 1 = 1 > 0$;
$1^3 - 3 \times 1 + 1 = 1 < 0$
The value of x that gives $x^3 - 3x + 1 = 0$ must lie between 0 and 1.

b $x_1 = 0.375$
$x_2 = 0.350\,911\,458\,3$
$x_3 = 0.347\,736\,944\,7$
$x_4 = 0.347\,349\,564\,4$
$x_5 = 0.347\,302\,774\,1$
$x = 0.3473$ (to 4 d.p.)

③ **a** $x^3 - 3x + 1 = 0 \rightarrow x = \frac{x^3}{3} + \frac{1}{3}$
$x^3 - 3x + 1 = 0$
Add $3x$ to both sides
$x^3 + 1 = 3x$
Divide both sides by 3
$\frac{x^3}{3} + \frac{1}{3} = x$

b $x_1 = 0.\dot{6}$

$x_2 = 0.432\,098\,765$

$x_3 = 0.360\,225\,625\,5$

$x_4 = 0.348\,914\,592\,7$

$x_5 = 0.347\,492\,449\,5$

$x = 0.35$ (to 2 d.p.)

c $(0.35)^3 - 3 \times 0.35 + 1 = -0.007\,125$

The solution satisfies the equation accurate to 1 d.p.

Unit 4 Histograms with unequal class widths

AO1 Fluency check

① **a** $0.8 \times 2 = 1.6$ **b** $5 \times 1.4 = 7$

② **a** $12 \leqslant a < 15$ **b** $24 - 0 = 24$ months

c Median = 14.5th value; $12 \leqslant a < 15$

d 12.839 = 13 months.

Confidence questions

①

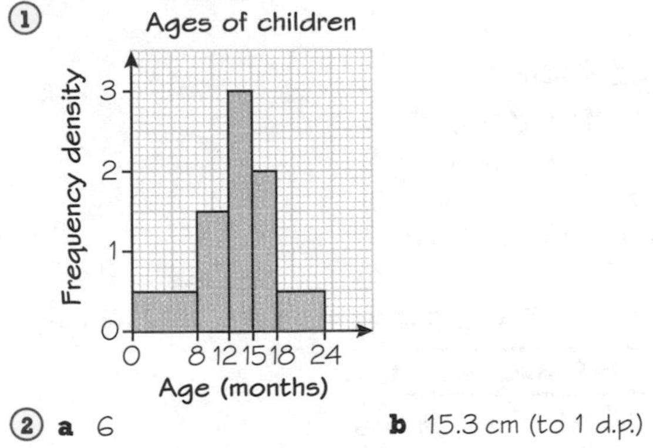

② **a** 6 **b** 15.3 cm (to 1 d.p.)

Skills boost 1 Drawing histograms

Guided practice

Time, t (seconds)	Frequency	Class width	Frequency density
$0 \leqslant t < 20$	2	$20 - 0 = 20$	$\frac{2}{20} = 0.1$
$20 \leqslant t < 25$	7	5	$\frac{7}{5} = 1.4$
$25 \leqslant t < 35$	12	10	$\frac{12}{10} = 1.2$
$35 \leqslant t < 40$	1	5	$\frac{1}{5} = 0.2$

① **a**

Length, l (mm)	Frequency	Class width	Frequency density
$5 \leqslant l < 10$	3	5	0.6
$10 \leqslant l < 12$	10	2	5
$12 \leqslant l < 15$	9	3	3
$15 \leqslant l < 20$	17	5	3.4

b, c

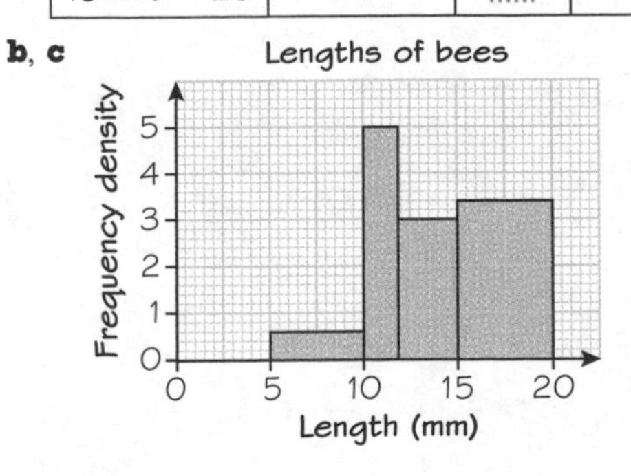

②

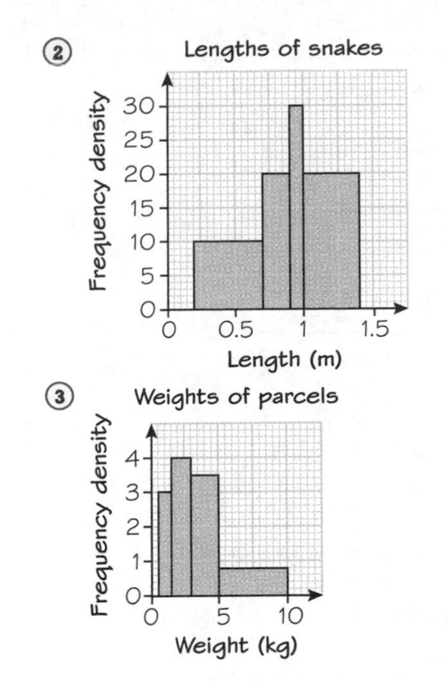

Skills boost 2 Interpreting histograms

Guided practice

a Class width = 10

Frequency density = 0.3

Area = 10 × 0.3 = 3

b Frequency density = 1.5

Area = 2 × 1.5 = 3

① **a**

Typing speed, t (wpm)	Number of people
$10 \leqslant t < 15$	4
$15 \leqslant t < 20$	5
$20 \leqslant t < 22$	5
$22 \leqslant t < 25$	6
$25 \leqslant t < 35$	2

b 22

② **a**

Height, h (cm)	Frequency
$65 \leqslant h < 67.5$	15
$67.5 \leqslant h < 70$	30
$70 \leqslant h < 72$	14
$72 \leqslant h < 76$	8

b 67

③ **a** 42 **b** 28 **c** 16

④ **a** 6 **b** $14 + 6 = 20$

c $30 + 14 + 6 = 50$

⑤ **a** 34 **b** 29

Skills boost 3 Estimating averages and range from a histogram

Guided practice

Height, h (cm)	Frequency
$10 \leqslant h < 15$	7
$15 \leqslant h < 18$	12
$18 \leqslant h < 21$	15
$21 \leqslant h < 24$	9
$24 \leqslant h < 30$	6

Median = $\frac{49 + 1}{2}$ = 25th value

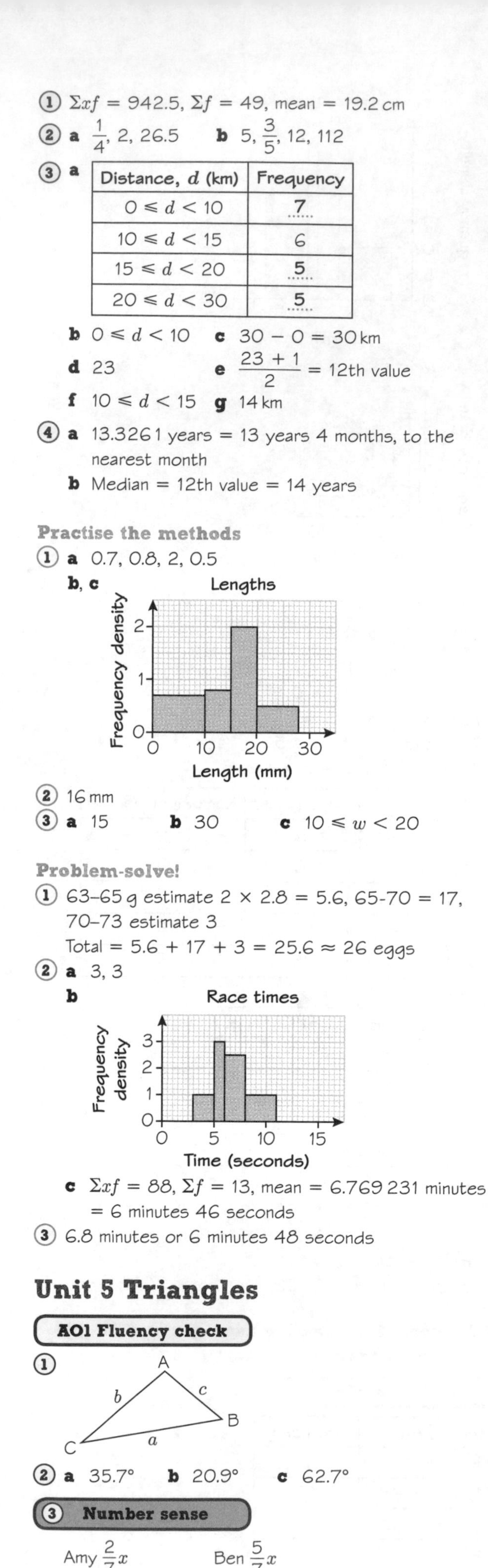

① $\Sigma xf = 942.5$, $\Sigma f = 49$, mean = 19.2 cm

② **a** $\frac{1}{4}$, 2, 26.5 **b** 5, $\frac{3}{5}$, 12, 112

③ **a**

Distance, d (km)	Frequency
$0 \leqslant d < 10$	7
$10 \leqslant d < 15$	6
$15 \leqslant d < 20$	5
$20 \leqslant d < 30$	5

b $0 \leqslant d < 10$ **c** $30 - 0 = 30$ km

d 23 **e** $\frac{23 + 1}{2} = 12$th value

f $10 \leqslant d < 15$ **g** 14 km

④ **a** 13.3261 years = 13 years 4 months, to the nearest month

b Median = 12th value = 14 years

Practise the methods

① **a** 0.7, 0.8, 2, 0.5

b, c

② 16 mm

③ **a** 15 **b** 30 **c** $10 \leqslant w < 20$

Problem-solve!

① 63–65 g estimate $2 \times 2.8 = 5.6$, 65-70 = 17, 70–73 estimate 3

Total = 5.6 + 17 + 3 = 25.6 ≈ 26 eggs

② **a** 3, 3

b

c $\Sigma xf = 88$, $\Sigma f = 13$, mean = 6.769 231 minutes = 6 minutes 46 seconds

③ 6.8 minutes or 6 minutes 48 seconds

Unit 5 Triangles

AO1 Fluency check

①

② **a** 35.7° **b** 20.9° **c** 62.7°

③ **Number sense**

Amy $\frac{2}{7}x$ Ben $\frac{5}{7}x$

Confidence questions

① 24.1 cm² ② 40°

③ 35.3° ④ (1.5, 3)

Skills boost 1 Calculating the area of a triangle

Guided practice

Area $= \frac{1}{2}xz\sin Y$

$= \frac{1}{2} \times 8 \times 5 \times \sin 40°$

$= 12.8557\ldots$

① **a** 24.2 cm² **b** 11.8 m²

② 46.2 cm³ or 46.3 cm³ depending on which sides and angles are used for calculation.

Skills boost 2 Finding angles and sides of a triangle when you know the area

Guided practice

$28 = \frac{1}{2} \times 7 \times 12 \times \sin P$

$28 = 42 \times \sin P$

① **a** 26.4° (to 1 d.p.)

b 27.0° (to 1 d.p.)

② **a** 7.4 cm **b** 61 mm

③ ∠R = 47.7°, $x = 66°$

Skills boost 3 Using the cosine rule to find an angle

Guided practice

$a^2 = b^2 + c^2 - 2bc\cos A$

$9^2 = 6^2 + 4^2 - 2 \times 6 \times 4 \times \cos A$

$81 = 52 - 48\cos A$

$29 = -48\cos A$

$48\cos A = -29$

$\cos A = \left(-\frac{29}{48}\right)$

$A = \cos^{-1}\left(-\frac{29}{48}\right) = 127.2°$ (to 1 d.p.)

① **a** 60.3° **b** 55.1°

② **a** $\cos A = \frac{b^2 + c^2 - a^2}{2bc}$

b 0.9617… **c** 15.9°

③ ∠M = 36°, ∠L = ∠N = 72°

Skills boost 4 Dividing a line in a given ratio

Guided practice

$AD = \frac{2}{3}AC = \frac{2}{3} \times 12 = 8$

$PD = \frac{2}{3}BC = \frac{2}{3} \times 9 = 6$

Work out the coordinates of P.

P(8, 9)

6

A (0, 3) 8 D(8, 3)

P = (8, 9)

① (6, 3) ② (0, −1)

Practise the methods

① **a** 20.7° **b** 30°

② 18.7 cm²

③ 13.9 cm
④ 33.6° (to 1 d.p.)
⑤ (−2, 0)

Problem-solve!

① 52.4 cm²
② **a** 36.8° (to 1 d.p.)
  b 12.6 cm²
③ $m = 3, n = 2$
④ **a** $6 = \frac{1}{2} \times 4 \times 6 \sin B$, $6 = 12 \sin B$,
  $\sin B = \frac{6}{12} = 0.5$
  b **i** 30° **ii** 150°
⑤ ∠M = 119°
⑥ B = (1, −2)

Unit 6 Pythagoras and trigonometry in 3D

AO1 Fluency check

① **a** $\sqrt{61} = 7.8$ cm (to 1 d.p.)
  b $\sqrt{40} = 6.3$ cm (to 1 d.p.)
② **a** 38.7° **b** 39.5° **c** 72.5°
③ 109.5° (to 1 d.p.)

Confidence questions

① 8.2 cm (to 1 d.p.)
② 34.9° (to 1 d.p.)
③ 21.5° (to 1 d.p.)

Skills boost 1 Using Pythagoras' theorem in 3D solids

Guided practice

$FH^2 = 3^2 + 7^2$
$FH = \sqrt{58}$ cm
$d^2 = 4^2 + (\sqrt{58})^2$
$= 16 + 58$
$d = \sqrt{74} = 8.6$ cm (to 1 d.p.)

① **a** 5 cm **b**

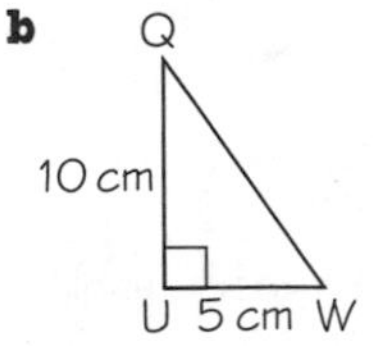

c 11.2 cm

② 8.3 cm
③ 10.8 cm

Skills boost 2 Using trigonometry in 3D solids

Guided practice

$4^2 = 5^2 + 5^2 - 2 \times 5 \times 5 \times \cos E$
$16 = 50 - 50 \cos E$
$\cos E = \frac{50 - 16}{50} = \frac{34}{50}$
$E = \cos^{-1}\left(\frac{34}{50}\right) = 47.2°$ (to 1 d.p.)

① 6.1 cm
② **a** 86.3° **b** 7.2 cm² (to 1 d.p.)
  c 60.5 cm³ (to 1 d.p.)
③ 61.9° (to 1 d.p.)
④ 10.5 cm

Skills boost 3 Finding the angle between a line and a plane

Guided practice

$AC^2 = 10^2 + 10^2$
$AC^2 = 200$
$AC = \sqrt{200}$
$\cos x = \frac{\frac{1}{2}\sqrt{200}}{12}$
$x = \cos^{-1} \frac{\frac{1}{2}\sqrt{200}}{12} = 53.9°$ (to 1 d.p.)

① **a** 8^2, 4^2, $\sqrt{80}$ cm **b** $\sqrt{80}$, 33.9°
② 65.4° (to 1 d.p.)

Practise the methods

① **a**

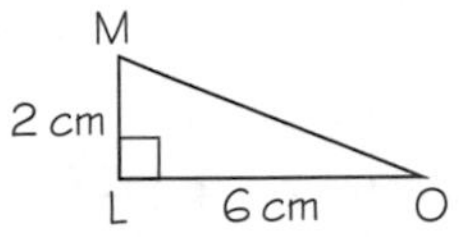

$MO = \sqrt{40}$
$= 6.3$ cm (to 1 d.p.)

  b

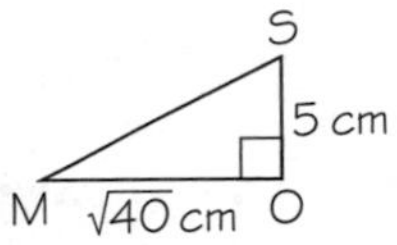

② 8.1 cm (to 1 d.p.)
③ 34.7° (to 1 d.p.)
④ **a** 50° **b** 65°
⑤ 38.3°

Problem-solve!

① **a** 60.9° **b** 4.4 cm
② **a** 36.5 cm² (to 1 d.p.)
  b 53.5°
③ **a** 11.5 cm **b** 20.3°
④ 25.3°
⑤ **a** 35.3°
  b All cubes are similar, so corresponding angles within them are equal.

Unit 7 3D geometry

AO1 Fluency check

① **a** 200π cm³ **b** $\frac{1372}{3}\pi$ cm³
② **a** 2 **b** 8 **c** 224 cm³

Confidence questions

① 79.6 cm³ (to 1 d.p.) ② 47.8 cm³ (to 1 d.p.)

Skills boost 1 Finding volumes and surface areas of solids

Guided practice

Volume of cone A $= \frac{1}{3}\pi r^2 h$
$= \frac{1}{3}\pi \times 8^2 \times 10 = \frac{640}{3}\pi$ cm³
Volume of cone B $= \frac{1}{3}\pi r^2 h$
$= \frac{1}{3}\pi \times 4^2 \times 5 = \frac{80}{3}\pi$ cm³
Volume of frustum = volume of cone A − volume of cone B
$= \frac{640}{3}\pi - \frac{80}{3}\pi = \frac{560}{3}\pi$ cm³

① **a** $400\pi\ \text{cm}^3$ **b** $50\pi\ \text{cm}^3$ **c** $350\pi\ \text{cm}^3$

② $26\pi = 81.7\ \text{cm}^3$

③ **a** $128\ \text{cm}^3$ **b** 10.0 cm (to 1 d.p.)

④ Volume of smaller cone $= 72\pi\ \text{cm}^3$,
volume of frustum $= \frac{296}{3}\pi\ \text{cm}^3$; the frustum has the greater volume.

⑤ Volume of cone $= \frac{15}{2}\pi$, volume of hemisphere $= \frac{9}{4}\pi$
Volume of plastic $= \frac{21}{4}\pi \approx 16\ \text{cm}^3$

⑥ Volume of cone $= 8\pi\ \text{cm}^3$
Volume of hemisphere $= \frac{16}{3}\pi\ \text{cm}^3$
Total volume $= \frac{40}{3}\pi\ \text{cm}^3$
Mass $= \frac{40}{3}\pi \times 2.7 = 36\pi = 113.1$ grams

⑦ **a** $l = 18.973\ldots$
Curved surface area $= 357.6\ \text{cm}^2$
b Curved surface area of small cone $= 248.3647\ldots\ \text{cm}^2$
Curved surface area of frustum $= 109.280\ldots\ \text{cm}^2$
Total surface area $= 300.917\ldots = 301\ \text{cm}^2$ (to 3 s.f.)

Skills boost 2 Finding volumes and surface areas of similar solids

Guided practice

$k = \sqrt[3]{27} = 3$
$k^2 = 3^2 = 9$
Surface area of A $= k^2 \times 39$
$= 9 \times 39 = 351\ \text{cm}^2$

① **a** $k^3 = 8$
b $k = 2$
c $k^2 = 4$
d $320 \times 4 = 1280\ \text{cm}^2$

② $800\ \text{cm}^2$

③ **a** $\frac{1}{2}$ **b** $\frac{1}{4}$ **c** $49\ \text{cm}^2$

④ $60\ \text{cm}^2$

⑤ **a** 2 **b** 8 **c** $304\ \text{cm}^3$

⑥ $2.22\ \text{cm}^3$ (to 2 d.p.)

⑦ **a** 1.5 **b** $(\sqrt{1.5})^3 = 1.8371\ldots$ or $\frac{3\sqrt{6}}{4}$
c $367.4\ \text{cm}^3$ (to 1 d.p.)

⑧ $86.0\ \text{mm}^3$ (to 1 d.p.)

⑨ **a** $\frac{56}{98} = \frac{4}{7}$ **b** 0.6886...
c $56.5\ \text{cm}^2$ (to 1 d.p.)

⑩ $199.2\ \text{cm}^2$ (to 1 d.p.)

Practise the methods

① **a** cone B
b Surface area of B $= \frac{8}{5} \times$ surface area of A
c **i** $\sqrt{\frac{8}{5}} = \frac{2\sqrt{10}}{5}$ or 1.2649...
ii $\frac{16\sqrt{10}}{25} = 2.0238\ldots$

② **a** $122.5\ \text{cm}^3$ (to 1 d.p.)
b Curved surface area $= 113.097\ldots$
Total surface area $= 144.5\ \text{cm}^2$ (to 1 d.p.)

③ $1400\pi\ \text{cm}^3$

Problem-solve!

① **a** 7.5 cm **b** 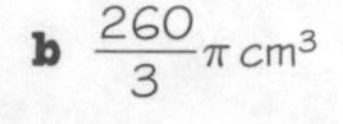 $\frac{260}{3}\pi\ \text{cm}^3$

② $1290\ \text{cm}^3$

③ 29.0 cm

④ $0.185\ \text{m}^3 = 185$ litres

⑤ $11\,160\ \text{cm}^3$ (to 4 s.f.)

Unit 8 Algebraic and geometric proof

AO1 Fluency check

① **a** $\angle C = 40°$ and $\angle D = 60°$ (corresponding angles). $\angle A$ is common angle. $\triangle ABE$ and $\triangle ACD$ have same angles, so are similar.
b Scale factor $= \frac{AC}{AB} = \frac{9}{6} = \frac{3}{2} = 1.5$
$CD = 1.5 \times 5 = 7.5$ cm

② **Number sense**

a $2n - 1$, $2n + 1$ **b** $2n$, $2(n + 1)$

Confidence questions

① $(2n + 1)(2m + 1) = 4mn + 2m + 2n + 1$
$= 2(2mn + m + n) + 1 = 2k + 1$, which is odd.

② Right-angled isosceles triangles all have angles 90°, 45°, 45°.

As they all have the same angles, they are all similar.

③

AB = AD (properties of symmetrical
DC = CB arrowhead, given)
AC is common side.
$\triangle ADC$ is congruent to $\triangle ABC$ (SSS).

④

$x + y = 180°$ (opposite angles in cyclic quadrilateral)
$z + y = 180°$ (angles on straight line)
Therefore $z = x$

Skills boost 1 Proving results about odd numbers, even numbers and squares

Guided practice

$(2n)^2$
$= 4n^2$

$(2n + 2)^2$
$= 4n^2 + 8n + 4$

$(2n)^2 + (2n + 2)^2 = 4n^2 + 4n^2 + 8n + 4$
$= 8n^2 + 8n + 4$
$= 4(2n^2 + 2n + 1)$

① $4(2n + 1)$ is a multiple of 4.

② $n - 1 + n + n + 1 = 3n$

③ $2m + 2n = 2(m + n)$, which is a multiple of 2, so even.

④ $(2n + 1) - (2m + 1) = 2n - 2m = 2(n - m)$

⑤ $(2n - 1)^2 = 4n^2 - 4n + 1$
$(2n + 1)^2 = 4n^2 + 4n + 1$
Sum $= 8n^2 + 2 = 2(4n^2 + 1)$, which is a multiple of 2, so even.

⑥ **a** 32 **b** $(2n + 1)^2 - (2n - 1)^2 = 8n$

Skills boost 2 Proving geometric results using similarity

Guided practice

∠ACB = ∠AED (corresponding angles in parallel lines)
AC : CE = 2 : 1

$\frac{AC}{AE} = \frac{2}{2 + 1} = \frac{2}{3}$

Work out the area scale factor.

$\left(\frac{2}{3}\right)^2 = \frac{4}{9}$

① **a** All angles 60°, $\frac{AC}{DF} = \frac{BC}{EF} = \frac{AB}{DE} = \frac{3}{5}$

b All angles 60°, $\frac{GH}{JK} = \frac{HI}{KL} = \frac{GI}{JL} = \frac{x}{y}$
Corresponding sides in same ratio.
OR enlargement from GHI to JKL scale factor $\frac{y}{x}$
y and x can take *any* values.

② **a** All angles 90°. Sides in ratio $\frac{2}{4}$

b All angles 90°. Corresponding sides in same ratio $\frac{y}{x}$ or $\frac{x}{y}$. y and x can take *any* values.

③ **a** ∠D = ∠B = 52° (alternate angles)
∠DCE = ∠ACB = 61° (vertically opposite angles)
∠E = ∠A = 67° (alternate angles, angle sum of a triangle)
Both triangles have the same angles, so are similar.

b ∠V = ∠Z = z (alternate angles)
∠W = ∠Y = y (alternate angles)
∠YXZ = ∠VXW = x (vertically opposite angles)
Both triangles have the same angles, so are similar.

c Two intersecting lines between two parallel lines always create two triangles. The result in part **b** is true for all angles x, y, z, and x, y, and z can take any values that sum to 180°.

④ **a** ∠V common angle, ∠VZW = ∠VYX (corresponding angles in parallel lines), ∠VWZ = ∠VXY (corresponding angles in parallel lines)

b VZ : ZY = 1 : 3

1 3
V Z Y

$\frac{VZ}{VY} = \frac{1}{1 + 3} = \frac{1}{4}$, VZ $= \frac{1}{4}$VY

c Area scale factor $= \left(\frac{1}{4}\right)^2 = \frac{1}{16}$

Therefore area of ΔVZW $= \frac{1}{16}$ area of ΔVYX.

Skills boost 3 Proving geometric results using congruence

Guided practice

OA = OC (radii)
∠OBA = ∠OBC = 90° (given)

① Prove that ∠YWX = ∠ZWX = 90°.
So triangle XYW and triangle XZW are congruent
XW bisects YZ, XY = XZ, YW = WZ.
YW = XW (common)
XY = YZ (isosceles triangle)
YW = WZ (given)
so ΔXYW = ΔXZW (SSS)
∠YWX + ∠ZWX = 180°, so ∠YWX = ∠ZWX = 90°

②
Q
P T R
S

Prove that ∠PTQ = ∠PTS = 90°
PQ = PS (property of kite, given)
QT = TS (PR bisects QS, given)
PT is common side.
So ΔPQT is congruent to ΔPST (SSS).
Therefore ∠PTQ = ∠PTS and ∠PTQ + ∠PTS = 180°
So ∠PTQ = ∠PTS = 90° and PR is perpendicular to QS.

Skills boost 4 Proving circle theorems

Guided practice

OB = OC = OA = OD (radii)
∠OBC = ∠OCB = y (base angles in isosceles triangle)
∠DOA = $2x$ (exterior angle = sum of interior opposite angles)
So ∠AOC $= 2x + 2y = 2(x + y)$
∠ABC $= x + y$

① ∠XOZ = 180° (angles on a straight line)
∠XOZ = 2∠XYZ (angle at the centre is twice the angle at the circumference)
∠XYZ = 90°

② **a**

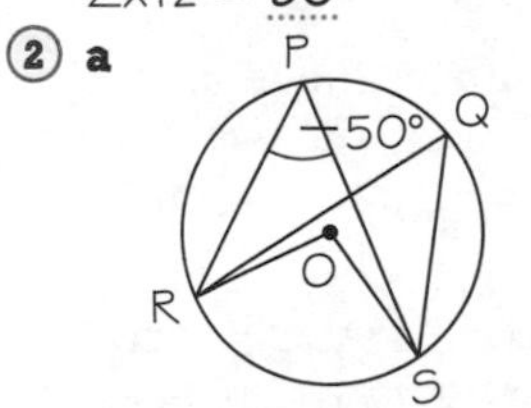

b ∠ROS = 100° (angle at the centre is twice the angle at the circumference)

c ∠ROS = 2∠RQS = 100° (angle at the centre is twice the angle at the circumference), ∠RQS = 50°

③ ∠AOD = 2∠ABD = $2x$ (angle at the centre is twice the angle at the circumference)
∠AOD = $2x$ = 2∠ACD (angle at the centre is twice the angle at the circumference)
Therefore ∠ACD = x
∠ABD = ∠ACD

Practise the methods

1. DE = GF, DG = EF (opposite sides of a parallelogram); DF is common; so $\triangle$DFG is congruent to $\triangle$DFE (SSS).
2. Let the even number be $2n$ and the odd number be $2m + 1$.
$2m + 1 - 2n = 2(m - n) + 1$
$= 2k + 1$, which is an odd number.
3. All angles in both rectangles are 90°. AE = FG, AG = EF. $\frac{AE}{AB} = \frac{2}{3} = \frac{GF}{DC}$ All sides are in the same ratio. AEGF is an enlargement of ABCD, scale factor $\frac{2}{3}$
4. 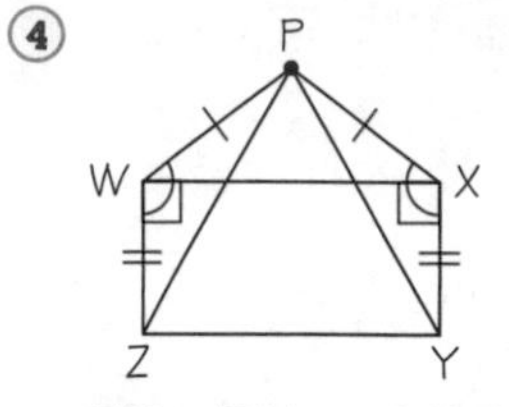

PW = PX (given), WZ = XY (opposite sides of rectangle, given), $\angle$PWX = $\angle$PXW (base angles in isosceles triangle), $\angle$ZWX = $\angle$YXW (90°, angles in rectangle). $\angle$ZWP = $\angle$PWX + $\angle$ZWX = $\angle$PXW + $\angle$YXW = $\angle$YXP. $\triangle$ZWP is congruent to $\triangle$YXP (SAS). Therefore PZ = PY
5. 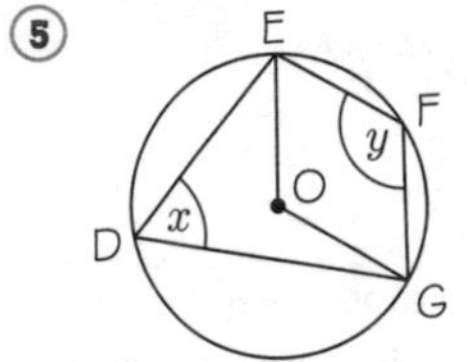

Obtuse $\angle$GOE = $2x$ (angle at the centre is twice the angle at circumference, $\angle$GDE),
Reflex $\angle$GOE = $2y$ (angle at the centre is twice the angle at circumference, $\angle$GFE), $2x + 2y = 360°$ (angles around a point), $2(x + y) = 360°$,
$x + y = 180°$

Problem-solve!

1. $2m + 1 - (2n + 1) = 2m - 2n = 2(m - n) = 2k$, which is even.
2. $2n(2m + 1) = 4nm + 2n = 2(2nm + n) = 2k$, which is even.
3. $\angle$CAB = $\angle$CED (alternate angles), $\angle$ABC = $\angle$EDC (alternate angles), $\angle$ACB = $\angle$DCE (vertically opposite angles). All angles in $\triangle$ACB and $\triangle$DCE are the same, so the two triangles are similar.
4. **a** 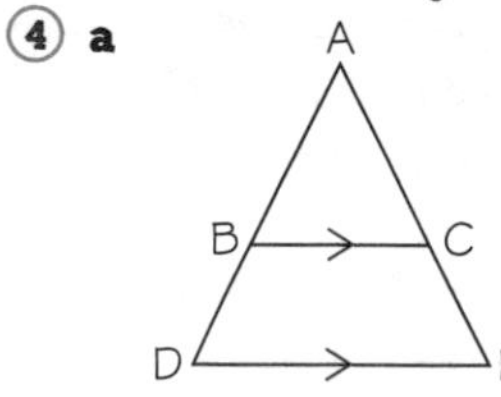

b $\angle$A is common angle, $\angle$B = $\angle$D and $\angle$C = $\angle$E (corresponding angles)
c Linear scale factor = ratio of heights
$\frac{\text{height of top cone}}{\text{height of whole cone}} = \frac{a}{a + b}$

5. **a** $\angle$OAC = $\angle$OBC = 90° (tangent meets radius at 90°)
b OA = OB (radii of circle, given), OC is common side, so $\triangle$OAC is congruent to $\triangle$OBC (RHS).
c As $\triangle$OAC is congruent to $\triangle$OBC, the two corresponding sides AC and CB are equal, therefore tangents from a point to a circle are equal.

Unit 9 Circles

AO1 Fluency check

1. $y = -\frac{1}{5}x + 4$
2. 5π cm²
3. 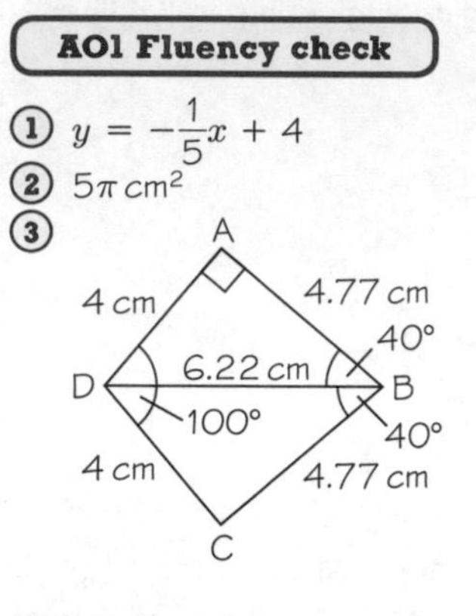

Confidence questions

1. 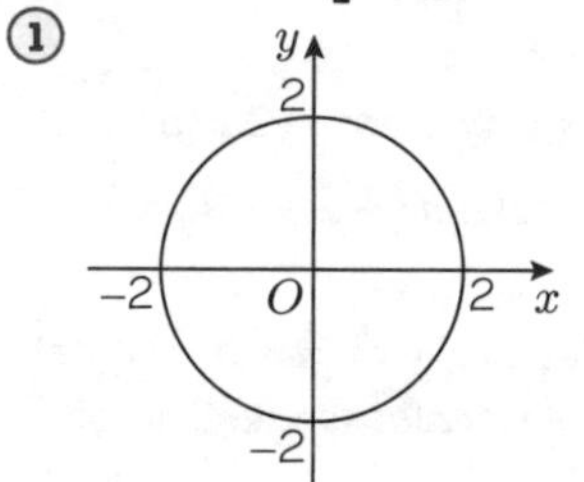

2. $\angle$COB = 40° (angle at centre is twice angle at circumference)
$\angle$OCB = 90° (tangent is perpendicular to radius)
$x = 50°$ (angles in triangle sum to 180°)
3. $y = \frac{4}{3}x - \frac{25}{3}$ or $3y - 4x = -25$
4. 2.5 cm² (to 1 d.p.)

Skills boost 1 Drawing graphs of circles

Guided practice

$r^2 = 81$

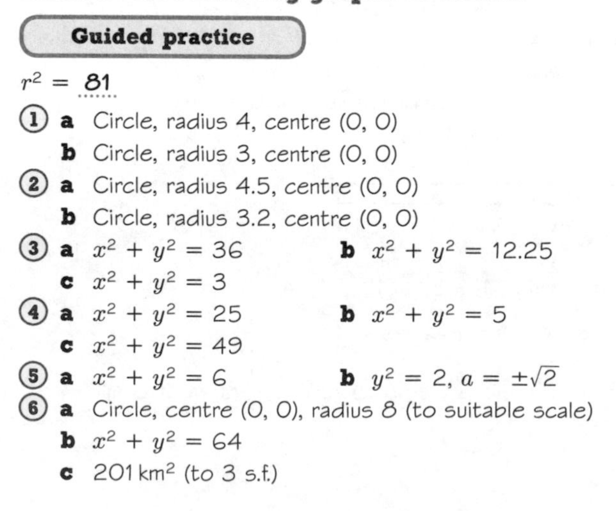

1. **a** Circle, radius 4, centre (0, 0)
b Circle, radius 3, centre (0, 0)
2. **a** Circle, radius 4.5, centre (0, 0)
b Circle, radius 3.2, centre (0, 0)
3. **a** $x^2 + y^2 = 36$ **b** $x^2 + y^2 = 12.25$
c $x^2 + y^2 = 3$
4. **a** $x^2 + y^2 = 25$ **b** $x^2 + y^2 = 5$
c $x^2 + y^2 = 49$
5. **a** $x^2 + y^2 = 6$ **b** $y^2 = 2, a = \pm\sqrt{2}$
6. **a** Circle, centre (0, 0), radius 8 (to suitable scale)
b $x^2 + y^2 = 64$
c 201 km² (to 3 s.f.)

Skills boost 2 Finding angles and lengths in tangent diagrams

Guided practice

PQ = PR (tangents from a point)
OQ = OR (radii)
∠OQP = ∠ORP = 90° (tangent is perpendicular to radius)

a ∠ROQ + 90° + 90° + 40° = 360° (angles in a quadrilateral sum to 360°)

b OQ = 6 tan 20°

(1) **a** ∠OBA = ∠OCA = 90° (tangent is perpendicular to radius)
b ∠BAC = 60° (angles in a quadrilateral sum to 360°)

(2) ∠PSO = 90° (tangent is perpendicular to radius)
$4^2 + PS^2 = 10^2$ (Pythagoras' theorem)
PS = 9.2 cm (to 1 d.p.)

(3) **a** ∠YOZ = 130° (angles in a quadrilateral sum to 360°)
b ∠OYZ = 25° (base angles in isosceles triangle are equal)

(4) **a** ∠ONM = ∠OPM = 90° (tangent is perpendicular to radius)
∠NMP = 70° (angles in a quadrilateral sum to 360°)
b MO is a line of symmetry of the kite MNOP.
c $\sin 35° = \frac{5}{MO}$, MO = 8.7 cm (to 1 d.p.)
or $\cos 55° = \frac{5}{MO}$, MO = 8.7 cm (to 1 d.p.)

(5) ∠OED = ∠OFD = 90° (tangent is perpendicular to radius)
OED is a right-angled triangle.

E
7 cm
75°
O D

(OD is a line of symmetry of the kite OEDF)
$\cos 75° = \frac{7}{OD}$, OD = 27.0 cm (to 1 d.p.)

Skills boost 3 Finding equations of radii and tangents

Guided practice

a Gradient of radius $= \frac{12}{-5} = -\frac{12}{5}$

Use $y = mx + c$

$y = -\frac{12}{5}x + c$

b $12 = \frac{5}{12} \times -5 + c$

$c = 12 + \frac{25}{12} = \frac{169}{12}$

(1) **a** $\frac{5}{12}$ **b** $y = \frac{5}{12}x$

(2) **a** **i** $-\frac{12}{5}$ **ii** $y = -\frac{12}{5}x$

b **i** $\frac{12}{5}$ **ii** $y = \frac{12}{5}x$

(3) **a** $y = \frac{4}{3}x$ **b** $-\frac{3}{4}$ **c** $y = -\frac{3}{4}x - \frac{25}{4}$

(4) **a** $y = \frac{8}{6}x = \frac{4}{3}x$

b $y = -\frac{3}{4}x + \frac{25}{2}$ or $y = -0.75x + 12.5$

(5) $-3x + 4y = 75$

(6) **a** OP: $y = -\frac{4}{3}x$
OS: $y = \frac{3}{4}x$

b The gradient of OP is the negative reciprocal of the gradient of OS, so the two radii are perpendicular.

c PT: $y = \frac{3}{4}x + 12.5$
ST: $y = -\frac{4}{3}x - \frac{50}{3}$

d T(−14, 2)

(7) $8y + 15x = 289$

Skills boost 4 Finding the area of a segment

Guided practice

Area of sector OAB $= \frac{50}{360} \times \pi \times 5^2$

Area △AOB $= \frac{1}{2} \times 5 \times 5 \times \sin 50°$

Area of segment = area of sector − area of triangle
$= \frac{125}{36}\pi - 9.5755...$

(1) **a** $\frac{49}{6}\pi$ cm² **b** 21.2176 cm²
c 4.44 cm² (to 2 d.p.)

(2) **a** 112.9°
b 4.72 cm² (to 2 d.p.)

Practise the methods

(1) **b** Gradient $= \frac{6}{8} = \frac{3}{4}$, $y = \frac{3}{4}x$ **c** $-\frac{4}{3}$

d $y = -\frac{4}{3}x + \frac{50}{3}$

(2) $x^2 + y^2 = 7$

(3) Area of sector $= \frac{245}{18}\pi$ cm²
Area of triangle = 24.1277… cm²
Area of segment = 18.6 cm² (to 1 d.p.)

Problem-solve!

(1) **a**

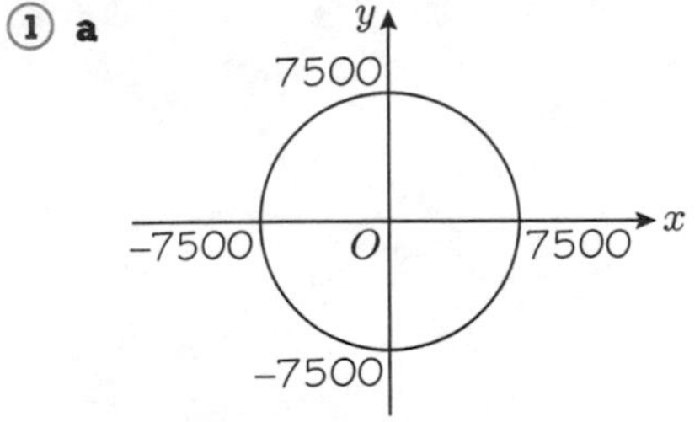

b $x^2 + y^2 = 56\,250\,000$

(2) XC = 4 cm, AY = 7 cm (tangents from a point are equal), CZ + ZB = 32 − 22 = 10 cm, CZ = 5 cm.

(3) ∠TUO = ∠TVO = 90° (tangent is perpendicular to a radius), ∠UOV = 135° (angles in quadrilateral sum to 360°)
Area of △OUV $= \frac{1}{2} \times 5 \times 5 \times \sin 135° = 8.84$ cm² (2 d.p.)

(4) Area of triangle $= \frac{1}{2}bh = \frac{1}{2} \times 2r \times r = 36$,
so $r = 6$
Area of semicircle $= \frac{1}{2}\pi r^2 = 18\pi$
Shaded area $= 18\pi - 36 = 18(\pi - 2)$ cm²